THE WORLD ACCORDING TO twitter

The World According to Twitter

David Pogue
and His 500,000 Followers

Black Dog
& Leventhal
Publishers
New York

Published by
Black Dog & Leventhal Publishers, Inc.
151 West 19th Street
New York, NY 10011
Distributed by
Workman Publishing Company
225 Varick Street
New York, NY 10014

Manufactured in the United States of America

Cover and interior design by Kevin Baier
Cover photograph: ©Sailorr/Shutterstock

ISBN-13: 978-1-57912-827-2
h g f e d c b a
Library of Congress Cataloging-in-Publication Data available on file.

Contents

Contents

Introduction

"WHY SHOULD I CARE WHAT YOU HAD FOR BREAKFAST?"

ARRRRGHH!!! I can't *stand* it when people say this about Twitter!

OK, asking that question is a fairly normal reaction when someone first hears about Twitter. I admit it: That's the reaction *I* had when I first heard about Twitter.

But that is *not* what Twitter is about—even if the little box you type into is labeled, "What are you doing?"

ABOUT TWITTER

Now, let's get one thing clear upfront: This is not a book *about* Twitter.

Twitter just happens to be the first communications channel in history that could have made possible a book like this: real-time, interactive collaboration with half a million smart, funny coauthors. What you'll find on these pages is their wit and wisdom, with very few references to Twitter itself.

But, in case you care, Twitter.com is a Web site where you can broadcast very short messages—140 characters, max—to anyone who's signed up to receive them. It's like a cross between a blog and a chat room.

Your "followers" might include six friends from high school, or, if you're movie star Ashton Kutcher, 2.5 million faithful fans. At the same time, you can sign up to follow the "tweets" of *other* people, which scroll up your screen in a big column, all mixed together, like the screenplay for a global cocktail-party scene.

Fortunately, most people do *not* broadcast the mundane details of their lives. Instead, they make wry observations. They send links to interesting stuff they've found online. They pass along breaking news. (All kinds of news breaks on Twitter before it hits the mainstream media: Barack Obama's choice of a running mate, the plane landing in the Hudson River, the Mumbai earthquake, the rebellions in Iran, and so on.)

And, oh yeah—they ask questions.

ABOUT MY CONVERSION

I'll be the first to admit that I was a Twitter skeptic. Like most first-timers, I found Twitter to be filled with confusing conventions, rules, and shorthand. The first time I covered Twitter for the *New York Times*, I wrote, "Like the world needs *another* ego-massaging, social-networking time drain?"

But that's the thing about Twitter: it takes a week or so to get it. And the possibilities slowly began to blow my mind.

My transformation into a full-blown Twitterphile began the day I served as a judge on a grant committee. I watched as a fellow judge asked his Twitter followers if a certain project had been tried before. In thirty seconds, his followers replied with Web links to the information he needed. I was amazed at the quality and quantity of the responses.

So a week later, I conducted a similar experiment myself. I was giving a talk in Las Vegas, and trying to explain Twitter by demonstrating it. As the audience watched, I fired up Twitter on the big screen and typed to my 5,000 followers: "I need a cure for hiccups . . . right now!"

Within seconds, responses came pouring in from all over the world. They were weird, wonderful, funny:

- Take 9 sips of water then say, "January." Laugh now, but you'll thank me when the hiccups are gone!
- Peanut butter on a spoon.
- Check your 401K. That should scare the hiccups right out of ya.
- I take large sips of bourbon. It doesn't stop the hiccups, but I stop caring!

We were all cracking up. This wasn't like anything else on the Internet. No Web page, chat room, or e-mail could have achieved the same effect, in real time, with this many participants. (The results of the hiccup-cure experiment are immortalized on page 74.)

It was rapidly becoming apparent that people on Twitter are not the same people who populate MySpace, Facebook, or YouTube. (Have you *seen* the level of discourse on YouTube? "Your such a tard" . . . "no U are!")

Twitterers are a different breed altogether. According to studies by PEW and Quantcast, we're an older, better-educated, higher-earning group. About 80 percent of us are over 25, and two-thirds of us have college degrees.

The bummer was that my followers' brilliance was trapped. Not just on Twitter—*on my screen*. With certain exceptions, when you get a reply on Twitter, you're the only person who sees it.

When I got home from that speaking trip, I mentioned that problem to my wife, Jennifer.

"You know what you should do," she said. "You should ask a question every night, and then publish the best answers in a book."

Now, Jennifer has always been a walking idea machine—her new-business ideas could fill a Wharton School catalog. But this one hit me right between the eyes.

This, I thought, could be a *really* interesting social experiment—and a darned entertaining book.

ABOUT THIS BOOK

And so it began. Every night, at about 11 p.m., I posed a new question to my followers on Twitter. "What's your million-dollar idea?" "What's your strangest habit?" "Make up a clever title for a sequel to a famous movie." (The time—11 p.m.—was designed to maximize the number of people who'd be on Twitter: night-owls on the East Coast, after-dinner tweeters on the West Coast, and maybe even some early risers in Europe.)

Then, for the next two hours, I'd sit there on the couch, reading the replies and cackling like a deranged person.

But even then, nobody could see these responses except me. After a couple of nights, I couldn't resist: I began *retweeting* five of the best responses. That means rebroadcasting them, passing them on to my entire group of followers, so that everyone could see them.

I used the standard retweet format, where you begin the message by crediting the original author, like this:

RT @justinchambers: "Snakes in the Terminal"

(RT means "retweet," and the @ symbol denotes a Twitter name. For example, my address on Twitter is @pogue.)

I've never had so much fun putting together a book. Apparently, the Twitterverse public also enjoyed the ride; during the three months of Twitter-booking, I somehow picked up 495,000 more followers.

Not everyone was happy, however. Bloggers, who can always be counted on for snarky reactions to anything, were quick to pile on.

Two things seemed to bug them. First, that I'd be taking credit for a book that I didn't actually write. "Apparently the days of actually 'writing' a 'book' are slowly coming to an end," wrote one blogger. "You know, craft, art, substance, the actual minutiae that all go into making a book a piece of work."

Second, the bloggers feared that I'd get rich off my followers' brilliance: "If you, lucky you, end up being selected to be a part of Pogue's scam project, you get compensated, right? Of course you do. Per Pogue himself, he'll send you 'a free copy of the book, inscribed to you.' Oh, wow, lucky day!"

Hmm. Well, on the first point, compilations are nothing new; consider Zagat restaurant guides, Bartlett's *Quotations*, joke books, household-hints books. As long as the micro-contributors all participated willingly, I didn't see the problem.

On the compensation point—well, this book's contributors knew the terms of the arrangement from the outset: If I include one of your tweets, you get an autographed copy of the book. And, as one participant noted, "I wrote 1 sentence and got a $13 book. That works out to $4,680 an hour. I'm OK with that."

In the end, I posed 95 questions. They generated over 25,000 responses. It took my editor and me weeks to winnow them down to the 2,524 winning tweets in this book. A book like this is typically shelved in the Humor section of the bookstore, but to my delight, it evolved into something that goes way beyond jokes. Here are life stories, greatest regrets, poems, brilliant inventions, advice for lovers, wry observations, hopes for the future, and words to live by...here's the whole world according to Twitter.

And a lot of great jokes.

As you'll soon see, this book's coauthors are some of the wittiest, sharpest, most interesting people on the Internet. They come from all over the world, they sleep in every conceivable time zone, they represent an astonishing range of life experiences.

And, thanks to the unique, real-time, communications channel called Twitter, they've successfully completed a massive collaborative experiment.

Too bad we'll never meet.

ABOUT THE TEAM

There was a lot more to this book than just asking questions and collecting the answers. The technical challenges were especially difficult.

First, tweets don't stay on the Internet forever. There is a search function on Twitter, but it "sees" only the last few days' worth of tweets. Therefore, just figuring out how to capture the thousands of responses to my questions became a daunting project.

Second, once we'd selected our winners, we had to *contact* them—to get their official permission to display their edited tweets in this book, and to find out their shipping addresses for the free books.

But we had only a single piece of information about each person: his or her Twitter name. No e-mail address, no phone number, no mailing address.

So what's the big deal? Just contact them on Twitter, right?

Sure, except that by the time we were ready to reach out to them, 234 of our winners had *changed* their Twitter names, or quit using Twitter altogether—or, in one case, died.

Fortunately, I had a team of three geniuses working with me.

First, my summer intern, University of Virginia student David Pierce, figured out how to capture all those incoming replies and massage all of them into a usable, editable format. Incredibly, he also managed to hunt down most of those 234 missing contributors. (He used sneaky tricks like Googling their Twitter names, on the premise that if you call yourself FirefighterNYC on Twitter, you might also be FirefighterNYC on Facebook or Gmail.) I'll never forget his fist-pumping "Yes!!" across the room each time he tracked down another one of the missing.

Second, we struck gold when we discovered Twitoaster.com. This Web site bestows Twitter with something that it otherwise lacks: *threading*. In other words, it groups replies with the tweets that inspired them. From his home near Paris, Arnaud Meunier, Twitoaster's creator, cheerfully offered to set up a custom Web site to help with this book project. At any time, we could click over to it to harvest the latest replies in a tidy spreadsheet format. Arnaud's genius saved us weeks of effort.

Finally, I enlisted the aid of consultant Geoff Coffey—author of *FileMaker Pro 10: The Missing Manual*—to work with our master FileMaker tweet database. He trained it to automate the massive permission-seeking task, generating private Twitter messages to our thousands of contributors. He also created a custom Web site for the book's winners, where they could approve my edits, grant permission for publication, and supply their contact information.

And while I'm thanking people, I should mention Mary Ann Madden. I've never met her, but for decades, she ran the weekly "Competition" on the back page of *New York* magazine. Her ingenious wordplay challenges inspired some of this book's punnier questions.

My editor at Black Dog & Leventhal, Becky Koh, was sent from heaven. She grasped the project innately—we could practically complete each other's sentences when talking about it—and her humor, smarts, and tact had a huge impact on the shaping of the book. In fact, it seems as though everyone at Black Dog & Leventhal is scrappy, smart, and funny. That includes the big dog himself, J.P. Leventhal, whose superb taste inspired him to publish the book in the first place. And designer Kevin Baier, who gave these pages just the amount of whimsy and class they deserved. (Did you spot his flip-book margin movie?)

Twitter, the company, is still struggling to keep up with its unexpected popularity. Yet at various times, Twitter's Maggie Utgoff, Jenna Sampson, Jillian West, Santosh Jayaram, and cofounder Evan Williams took time out to help with this project in indispensable ways. I found it amazing that they were so cool about fielding my requests.

My agent, Jim Levine, jumped on the concept instantly and found the best possible home for it. Lesa Snider was not only my first Twitter teacher, but also my Webmistress for this book project. And my children, Kelly, Tia, and Jeffrey, were patient and forgiving during my prolonged absence from family life.

Above all, this book owes its existence to my wife, Jennifer. The whole thing was her brilliant idea, and it all would have collapsed without her love, effort, and support. As one of the modern-day greeting cards on page 38 puts it: "To my Tweetheart on our anniversary. After all these years, you still make me say OMG."

Compose the subject line of an email message you really, really don't want to read.

To my former sexual partners, as required by law —@markowitz

Re: What seems to have been your car —@pumpkinshirt

From: eHarmony. Subject: Your profile has been rejected. —@jadawa

I hate to do this via email... —@SusanEJacobsen

Fwd:Fwd:Fwd:Fw:Catz! lol —@danblondell

What happened in Vegas did NOT stay in Vegas —@jschechner

Your Dad is Now Following You on Twitter. —@CathleenRitt

From: Your Doctor. Subj: Good news, bad news... —@Baszma

Urgent notice to everyone who was at the hot tub party last Saturday! —@lavasusan

From: AT&T. Subject: Your international roaming charges —@kvijayraghavan

Hi! Remember me? I'm in town!! —@Stefaniya

Error in lab results —@ricksva

From: NIH. Subj: Important new information on link between computer usage and rapid-onset dementia —@maineone

Honey, you saved those tax papers from 1978, right? —@pumpkinshirt

From: Yale Office of Admissions. File Size: 2K —@perryan

We need to talk. Call me. —@_not_THAT_guy

From: Your Petsitter. Subject: Before you open the door when you get home... —@brianwolven

From: Your Publisher. Subject: Ha, good one! Could you send the real chapter now, please? —@Lookshelves

Your GM common stock —@scottmarkarian

Did you mean to hit Reply to All? —@Maggie_Dwyer

From: Your eldest kid. Subject: How do you get chocolate sauce out of the sofa? —@aymroos

Add 1 letter to a famous person's name; explain.

Sonny Bongo: An upbeat percussionist —@davenik

Johann Sebastian Beach: He's tanned and well-tempered but—don't fugue with THIS guy! —@rponto

Yo Yo Mad: Angry cellist —@eboychik

James Blond: A spy who has more fun —@jml407c

Sparis Hilton: Hotel heiress who, thankfully, stays out of the tabloids —@lizardrebel

Tronto: Sidekick of the Canadian Lone Ranger —@pumpkinshirt

Scole Porter: Composer of the immortal song, "I Get a Kick Out of Chew" —@rponto

Hands Christian Andersen: Touchy-feely children's writer —@eboychik

Robert E. Leek: Fought the Onion troops to a standstill —@pumpkinshirt

Sarah Paling: Alaska governor to install tanning bed to fix personal whiteness issue —@Styminator

Buckminster Fullear: Inventor of the geodesic Q-tip —@eboychik

Thomas Hardly: Mediocre British novelist —@dguinee

Crush Limbaugh: Outspoken right-wing wrestler —@robertgdaniel

Henry Fjord: Introduced the assembly line to Norway —@pumpkinshirt

Scooby Doob: Cheech & Chong's crime-fighting canine —@rponto

Queen Evictoria: The unforgiving landlady —@cloud64

Neil Farmstrong: That's one small seed for (a) man, one giant crop for mankind —@pumpkinshirt

Malcolm XY: Civil-rights activist, definitely male —@pixelshot

FU2: Iconic rock band unafraid to respond to haters —@noveldoctor

Sean Penne: Starchy, overcooked actor/activist —@sassone

The Telephant Man: Trunk-call specialist —@rponto

Nomam Chomsky: Controversial, but extremely polite, writer and thinker —@pumpkinshirt

Gringo Starr: Best drummer north of the border —@eboychik

Baitman: Dark knight of the sea —@john_cox

Elvish Presley: Middle Earth's latest rock sensation —@alitheiapsis

Add 1 letter to a famous person's name; explain.

Malice B. Toklas: The long-time enemy of Gertrude Stein —@pumpkinshirt

Time Rice: Highly processed movie scores in just 20 minutes —@lsmith1964

Henry Wrinkler: The Fonz is getting old —@christopherbmac

Seminem: Half-hearted hip-hop artist —@rkaika

ICE-TP: For a fresh, tingly clean —@calindrome

Arthur Fontzarelli: Known as The Fontz, he invented such classic '50s typefaces as Aaaayrial —@pumpkinshirt

Sweeney Toadd: Demon amphibian of Fleet Street —@rponto

Twill Shortz: Publishes the Sunday NY Times cross-stitch pattern —@Maggie_Dwyer

Frank Skinatra: Adult film star with mob ties (but, naturally, sans neckties) —@pumpkinshirt

Barack GoBama: President of the Alabama Football Fan Club —@kevinpshan

Edward R. Murrlow: Iconic news commentator possessing subtle vanilla and oak overtones —@rponto

Mr. MT: Sad, existentialist action hero of the '80s —@pumpkinshirt

Chez Guevara: Revolutionary Franco/Latino restaurant —@CanonDude

Thomas Hobbies: Life is just a bunch of nasty, brutish and short weekend projects —@livbem

Attila the Hunk: Conquered Gaul with charm and good looks —@richardhamilton

Carrie Sunderwood: Champion log splitter with a voice like an angel —@EShahan

Janeane GaROFLalo: Very funny woman —@interweave

Vladimir Puttin: Most famous miniature golfer in Russia —@JimFl

LOL Cool J: The Jolly Rapper

—@esvbbv

Simon Cowbell: Snarky music guy you can hear from a long way off —@noveldoctor

Felicia Day-O: Spunky banana boat actress —@CharPrincessa

B.B.B. King: Legendary bluesman and founder of Better Business Bureau —@robertgdaniel

RUN-DMCA: Hip-hop group devoted to defending copyright —@vidiot_

Kale-El: Superman's vegetarian cousin —@pumpkinshirt

Jeff Probist: Host of a reality TV show that, trust me, you are grateful you never saw —@myhelfy

Gal Pacino: The first Academy Award–winning cross-dresser —@hobbesdream

Charlston Heston: You'll pry these dance shoes off my cold, dead feet —@3x10to8mps

IraQ Glass: Host of This Mesopotamian Life —@gskull

Mr. TA: The teacher's assistant who pities the foo' —@arclite

Man O' Wart: Ugliest thoroughbred, three years running —@pumpkinshirt

Aeronsmith: Brand of ergonomic chairs for aging rockers —@VenetianBlond

William Shatnerd: Half-captain, half-Vulcan Starfleet officer —@dmkanter

Nine Inch Snails: Gastropodal heavy-mucus band; some acclaim for album, The Slip 'n' Slide —@rponto

Rebar McEntire: The country singer with a voice like concrete —@pumpkinshirt

Eminema: Colonic-obsessed rapper —@vidiot_

Summarize a famous book in 140 characters.

I love you, but family is FREAKING OUT. Wait, maybe we can elope. Oh...You're dead. Screw it. (Romeo and Juliet) —@stevenaverett

God meets a nice Jewish girl. People are still talking. (The Bible) —@UNnouncer

Mothballs in wardrobe cause English children to hallucinate, crossbreed humans + animals, start a war, kill a lion. (Lion, Witch & Wardrobe) —@Remi_T

Single girl, pounds to lose
Keeps a diary, guzzles booze
Full of humor, angst & wit
She ends up with a handsome Brit (Bridget Jones's Diary) —@tassoula

A (The Scarlet Letter) —@hriefs

War is peace? Ignorance is strength? Big Brother can't be right. I won't give in. They can't break me. Oh, I love Big Brother. (1984) —@carolcdt

Bread can really ruin your life. (Les Misérables) —@jenatesse

It's some random person you didn't think of, in some random cult you never heard of, and all the Christians get upset. (The Da Vinci Code) —@christian_major

Couple conspires. Wife develops OCD out of guilt. There are some crazy old ladies with a cooking show involved. (Macbeth) —@OMG_Ponies

2 + 2 = 5 (1984) —@fourfootflood

The shoe is mine; scrub your own floors. (Cinderella) —@katnagel

You can make it through anything if you don't lose your head. (A Tale of Two Cities) —@pumpkinshirt

There and back again. (The Odyssey) —@katnagel

Damn whale. (Moby-Dick) —@GlennF

Spice is the variety of life. (Dune) —@hose311

Happy setting, heart-tugging tragedy, road to recovery, happily ever after! (Anything written, produced or directed by Walt Disney) —@BerryLowman

Awesome, dinosaurs! Oh wait... (Jurassic Park) —@zcott

Kids do the darndest things. (Lord of the Flies) —@mehughes124

Spiders help to keep everything kosher. (Charlotte's Web) —@pumpkinshirt

Buyer's remorse. (Paradise Lost) —@cornedbeefgents

Life is a tale told by an idiot, full of impenetrable run-on sentences. (The Sound and the Fury) —@pumpkinshirt

No. no. no. no. no. no. no. Oh, all right. (Green Eggs and Ham) —@worldexplorer

What if it's not just a cold? (The Stand) —@pumpkinshirt

Revenge is a dish best served many years and 1,200 pages later. (The Count of Monte Cristo) —@dhaucke

I made a man from spare parts. He didn't like it. (Frankenstein) —@JimFl

Don't tick off the big guy. (The Bible) —@alechosterman

Beware of Dog. (Animal Farm) —@MaryHr

Ask around about the boss before you take a job. (Moby-Dick) —@pumpkinshirt

I'm having an old friend for dinner. (The Silence of the Lambs) —@leahey

Four funerals and a wedding. (The Lord of the Rings) —@frumpa

Damn Yankees! (Gone with the Wind) —@MacSmiley

Small boat. Big fish. (Jaws) —@chops893

Long way to return a ring. (The Lord of the Rings) —@dloehr

Don't eat meat. (The Jungle) —@Chrisbestwick

He was beautiful, so beautiful. All I could think or write about was his beautiful beauty. Oh, and he was a vampire. (Twilight) —@dhersam

Love means never realizing you're knee-deep in clichés. (Love Story) —@pumpkinshirt

I got drunk and high all the time and did lots of bad stuff. Or at least I want you to think I did. (A Million Little Pieces) —@kerri9494

When in doubt, add more butter. (The French Chef Cookbooks by Julia Child) —@pumpkinshirt

Old man, big fish, tired arms. (The Old Man and the Sea) —@JoeFranscella

42 (Hitchhiker's Guide to the Galaxy) —@laurenist

ABCDEFGHIJKLMNOPQRSTUVWXYZ. (Webster's Dictionary) —@Sandydca

Write tight. (The Elements of Style) —@TonyNoland

I'd feel better about life if my penis still worked. (The Sun Also Rises) —@xarker

Fail whale. (Moby-Dick) —@laurenist

A woman drags Dante through hell. (Dante's Inferno) —@igorlikesthings

Brevity is the soul of twit. (The World According to Twitter) —@dloehr

Take a common abbreviation and tell us what it *really* stands for.

BMW: Break My Windows! —@NickDow

OPEC: Oil Produces Easy Cash —@danblondell

NATO: Newly Antiquated Tsk-tsk Organization —@danblondell

NATO: No Action, Talk Only —@sombuak

NATO: North America Tells Others —@hose311

NRA: Neanderthals Requiring Ammunition —@howardbeech

MD: Money Doc; PhD: Poor Hungry Doc —@SulaymanF

TiVo: Tape Is Very Old! —@CitizenGeek

FORD: Effer Only Runs Downhill —@Joethewalrus

TARP: Taxpaying-Americans Robbery Project —@BerryLowman

DOT: Deliberately Obstructing Traffic —@MyraB

DELTA: Don't Expect Luggage to Arrive —@tatopuig

DSL: Damn Signal Lag! —@pumpkinshirt

EPCOT: Extremely Photographic Community of Tourists —@Narniaexpert

Take a common abbreviation and tell us what it *really* stands for.

GM: Got Money? —@arthurra

GM: Government Motors —@vdalal

GM: Gone, Mostly —@leahey

CBS: Cancels Best Shows (Moonlight) —@Zakiya_

GPA: Godawful Predictor of Achievement —@ophiesay

CIA: Covert International Assassins —@Zakiya_

NASA: Nerds Are So Awesome —@pumpkinshirt

IRS: Income Removal Specialists —@Sofajenkins

AARP: Army of Ancients Recruiting Program —@jo_snail007

NAFTA: No American Factories Taking Applications —@dstubb

NSA: No Such Agency —@osxgirl

CIA: Cover-ups in the Interest of America —@1objectivist

NRA: Neighbors Requiring Armor —@sgoodin

SAT: Secure a Tutor —@pumpkinshirt

MLB: My Loaded Bat —@blastedblog

IRS: Institutional Robbery and Stealing —@HeathWilkes

AIG: All Is Gone —@ColleenHawk

TWITTER: Typing What I'm Thinking to Everyone Reading —@kevinpshan

PORSCHE: Proof of Rich Spoiled Children Having Everything —@kcheriton

AIG: Arrogance Incompetence and Greed —@jakemates

FORD: Found Our Recession Dollars! —@bryanb

TSA: Tub Stacking Agency —@charterjet

TSA: Thousands Standing Around —@judygressel

TSA: Take Scissors Away —@SulaymanF

NBC: Now Betting on Conan —@pumpkinshirt

I, Pogue, was born at 9:09 on March 9, weighed 9 lbs 9 oz. What's your cool numerology story?

My parents' birthdays are 1/10 and 10/1, respectively—& their social security numbers are sequential (one right after the other)! —@rondfw

Almost didn't race my bike on 6/6/06; decided to anyway against better instincts—led to high-speed crash and 2 surgeries! —@canmoremd

My shoe size matched my age perfectly from age 9, shoe size 9 through age 14, shoe size 14. (Thankfully, the trend ended at 15.) —@kellycroy

We met 10/10/95, I proposed 10/10/01, married 10/10/08. Realized the coincidence only later! We will marry again 10/10/09 (church). —@Kirchberg

My great aunts were twins born different years. One near midnight, New Year's Eve, 1912; the other, minutes later, New Year's Day, 1913. —@thespicegirl

Got 4 stitches at age 4, 6 at age 6, and 9 at age 9...this could become problematic. —@timmymacdonald

Married on 9/11; kid's birthdays are 11/11, 11/12, 12/11. 7 miscarriages on random dates, so 11 appears magical. —@auldfatbroad

I was born on Feb. 28, my sister on Nov. 28, my cousin on Jan. 28, and my parents were married on July 28. —@gizmosachin

I was born on 11/11 at 1:11 a.m., weighing 5 lbs, 6 oz (5 + 6 = 11). K is the first letter in my name and the 11th letter of the alphabet! —@khong79

My son was born on 5/5/99, weighed 7 lbs, 7 oz, 22 inches, at 11:11. —@schampi1

The numbers of every house I've ever lived in add up to 8: Born at #44, then moved to #53, then #80. Bought my first home last year: #71. —@AllisonRH

My mom's birthday is 7/18; her maternal grandpa's birthday was 7/19. Mine is 8/18; my maternal grandpa's was 8/19. Both 60 years apart. —@eemathnut

I was born on my father's birthday, my wife was born on HER father's birthday, and our son was born on my birthday. How odd is that? —@mnrainman

Pogue Sez

Actually, the numerology coincidences revealed in my question is only part of the story.

I was born at 9:09 a.m. on March 9, 1963 (note that 6 + 3 = 9). I weighed 9 pounds, 9 ounces. My parents had been married 9 years.

I turned 9 years old in 1972, where, once again, 7+2=9.

I've owned two houses. The first street address was 207 (whose digits equal 9); the second, where I live now, is 18 (ditto).

And my favorite number, of course, is 7.

I, Pogue, was born at 9:09 on March 9, weighed 9 lbs 9 oz. What's your cool numerology story?

I was at a baseball game in Oakland years ago. Score was 5–5. Time was 5:55. Batter was #55. He was hit by the pitch and broke his arm. —@scottknaster

Grandma always said she would live to be over 100. She was born 10/9/1907 and died 11/9/2007—100 years and 1 month. She did it! —@kathyaddison

My wife Jean and I were married on 5/7/05. My dad recognized the date by giving his toast as a haiku. —@DavidBThomas

This year my birthday will be 20/09/2009 (unless, for some reason, I happen to be in the US at the time). —@FredTweetzsche

My mother, sister and grandmother were all born on December 26 (different years). —@kdern

My husband's great-grandfather: Born on Valentine's Day, 1875, and died Valentine's Day, 1975. His last name was Valentine. —@merful

My sister was born on 2/2, married on 8/8 and had her first child 4/4. —@TimothyArcher

Wife & I were both born on a 26th, so we changed our anniversary to a 26th to make it easier. Neither can now recall the real day. —@tpete

I was born on 9/25—exactly 9 months following Christmas. I realized this when I was about 11, and was immediately horrified. Nguh... —@leahey

My car was hit by a surly 17-year-old going the wrong way on a one-way street at 3:30 on 03/03/03 going 30... —@Marsunderground

My son was born 7/16/07 (each adds up to 7-7-7) at 11:50 (sum is 7) and was 10 lbs, 6oz (= 7). (Yes, 10 lbs!!) —@MidnightCrafter

I have 3 brothers, all born on 11/15. I'm the odd one out—7/8. —@dbrewer80221

My grandmother died exactly at noon on 4/4 in 1993. My mom (her daughter) died exactly at noon on 4/4 in 2004. —@DrKoob

On 8/8/80 I turned 8. 8/8/88 I turned 8+8. (Now everyone knows how old I am, dammit.) —@VenetianBlond

I was born at 7am, the 7th month, 7th day of week. Weighed 7lbs 7oz. My name is Steven, which is "seven" with a t added. I like the number 3! —@MonkeyTwits

My son was born on the 3rd of April, at 3:33 p.m. He was in room 333, and was the 3rd baby born that day. It was also Easter Sunday. —@dyfhid

Former boss of mine had triplet girls on 3/3/93 (and no, she wasn't induced!). —@grousehouse

Not exactly cool, but brother died in Room 345 on 6/7/89. —@pendrift

Mom is 2 years, 9 months and 24 days older than Dad. I'm 2 years, 9 months and 24 days older than my brother. —@notaclevername

I, Pogue, was born at 9:09 on March 9, weighed 9 lbs 9 oz. What's your cool numerology story?

I was born 23:23, my first younger sister 03:03, my younger brother 14:14...only my 2nd younger sister fell out of line 5:15.
—@mj232385

I was born on 8/8, married on 10/10, and had daughter on 1/11. (Easy to remember anniversaries around here...)
—@kellyaharmon

Hubby & I remarried (each other) on divorce anniv. One niece married on our original anniv. Another niece married on our 2nd anniv. —@dfoulks

Make up a concept for a new TV show that's probably doomed.

Colonoscopy High Def —@Wylieknowords

Everybody Loves Cheney —@Notsewfast

Trading MySpaces —@tecnocato

"1's & 0's": A reality dating show between male and female computer programmers —@christopherbmac

Survivor, Final Season: Cannibal Edition —@Dorothy_Jean

Frenemies —@plocke

Reference Librarians Gone Wild —@margaretmontet

Quaker Gangs

—@Wylieknowords

Arch Enemies: The wacky hijinks of dueling podiatrists in the Greater Pittsburgh area —@pumpkinshirt

Guantánamo! —@jappio

Extreme Makeover: Gas Station Restroom Edition
—@hughesviews

FOUND: the story of a plane that lands safely at LAX
—@ImageSpecialist

Make up a concept for a new TV show that's probably doomed.

How I Met Your Mailman —@mrngoitall

Mama'Bama: When the president invites his mother-in-law to live at the White House, hilarity ensues —@ajbezark

IT Swingers —@nickfranklin

Law & Order: DMV —@pastorwalters

Top Janitor —@TiVoUpgrade

Proctology 90210 —@thebandnork

Make up a greeting card for a modern situation.

So sorry your job has been outsourced overseas! —@jschechner

I realized what happened when I heard that fateful moan.
I'm really, really sorry that the toilet claimed your phone.
—@scotthartman

Sorry this birthday card is late. E-mail was down and Facebook didn't remind me. —@alechosterman

(From Hallmark's new series, Relaxing with the Recession): Sorry Your Think Tank Tanked. —@Wylieknowords

In lieu of a gift, we've recycled 1,000 early iPhones in your honor. —@nunncookchicago

Match.com brought us together, YouTube pushed us apart.
But no matter what you tweet, you'll always be in my heart.
—@SusanEJacobsen

Sorry that I returned your electric car uncharged... —@Alonsomex

We heard about your tragedy, but now you're coming back.
After one more Blue Screen of Death, you bought that brand-new Mac! —@Sandydca

I'm a Mac. You're a PC. Let's virtualize. —@cyeoh6

Happy 1 millionth follower! —@filjedi

So sorry Yoga and Kabbalah have failed you! Hope Mindfulness will help you. Get Well Soon!! —@kvetchguru

To the cute barista who has no real idea who I am: This Valentine's Day, I'd like thank you for the eye contact. —@pumpkinshirt

To my Tweetheart on our anniversary. After all these years, you still make me say OMG. —@mattboom

Congratulations on your Google ranking! —@sassyscorpio71

I'm sorry your house is a cellphone dead zone... —@JakeHurlbut

Sorry to hear about your recent DMCA takedown notice. —@gvegas864

Sorry to see you had to change your relationship status to "Single." —@CafeChatNoir

Sorry your reality show co-contestants voted you off! —@pumpkinshirt

What a shame your brand-new laptop is obsolete already! —@jschechner

We hope Congress likes your next business plan. Keep trying! —@justcombs

You had me at "Geotag." —@cyeoh6

I'm sorry to hear your upgrade to Blu-Ray has rendered your DVD collection inadequate. —@pumpkinshirt

Saw the video of you on YouTube. I still love you. —@MyCatIsOnFire

You've gone Platinum! The RIAA congratulates you on 1,000,000 illegal Pirate Bay downloads of your soon-to-be released CD. —@eboychik

I'm sorry to hear about the passing of your pet, but so excited about the clone! —@pumpkinshirt

Life's twists and turns, make some folks bitter.
Forgive me for, not friending you on Twitter. —@wotten1

Sorry your TiVo cut off the final performance on "American Idol" this week. Hope you find it online! —@ascottfalk

You're truly a thoughtful neighbor...when you're not freeloading on my wireless network. —@hriefs

Glad to hear it wasn't swine flu. Feel better soon! —@dances_w_vowels

So sorry to hear you were nude sunbathing when the Google Maps Camera Car drove by. —@rponto

So sorry to hear your pay has been capped at $500K! —@KGWillison

Congrats on your new 401k balance—you've got a long future ahead of you! —@DrewJazz

A move to the country? We'd bid an adieu.
Our GPS, sadly, will never find you! —@dances_w_vowels

What was your greatest achievement (besides your kids)?

In 6th grade, I stood up to a bully twice my size. Guy had tortured me for years. Changed everything. —@kellycroy

I hit coolest home run in my Little League's history. Liner thru SS's legs, then thru LF's legs. Relay throw over backstop. We won! —@JohnPHolmes3

I yanked a toddler from stepping into traffic while his mom struggled to read the Jamba Juice menu. —@davedotdean

I graduated from law school at 20, and soloed with an orchestra (I'm a violinist) in Carnegie Hall at 18. —@DominiqueVance

Going to grad school in my 40s and being the oldest by a decade—came out of it with some of my closest friends and new worldview. —@knotman

After an afternoon in NYC, took a Metro North train home. Heard conductor ask for someone who knew CPR—I ended up delivering a baby! —@eddiepro

I grew up living well below poverty level in Miss., then put myself through college and graduated debt-free with a 3.9 GPA. —@laura_jeanette

At 16 I saved an 8-year-old girl from drowning—she jumped into the deep end of the pool—private picnic party. Lifeguards on break. —@1MysteryGirl

Riding bicycle from SF to LA 18 months after emergency liver transplant (and 10-day coma)...no one the wiser unless I flashed my scar. —@bekSF

Finding my daughter's stolen guitar on eBay and recovering it by joining a sting operation with a team of off-duty DEA agents. —@merful

Rescued a baby who had crawled out of her house and into the street. About 10 years ago. Wonder where/how she is now. —@melissajt

Going from 2 packages of cigarettes a day to zero. Smoke-free for 10 years. —@bitfiddler

I took a group of teenagers to Bolivia, where we gave some kids a new bathroom for their school. New entry stairs, too. —@gigemlee

Saving kid's life 3,000 miles away. My daughter's friend on AIM said he'd taken too many pills. I called police in FL, who found him. —@susanchamplin

Earning PhD in nuclear engineering while growing family from 1 child to 4. —@jlconlin

Immigrated to the US, became a citizen, and worked hard to get my wife to come over after years living apart. Most painful years ever. —@BlueGromit

Losing 145 pounds and running a marathon. —@nieves111

Getting off welfare. Went back to school at 26 and became a chemist (daughter was 4). —@chemrat

Coming back alive from Vietnam! —@usafalcon

Learning to play the trumpet while in college. (Yes, it was in order to impress a girl! The girl's history, but I still play.) —@rondfw

What's your greatest regret?

Not telling Grandma "Thank you" for the $5 in my 1979 Christmas card. She died on 2/25/80, before I could see her. I was 15. —@GrumpyHerb

Hurting my dad during my rebellious teenage years. He didn't deserve the hell I put him through. —@fanfrkntastic

Letting my mom talk me into a wedding dress I didn't want—everything was perfect except the dress! —@TiffanyKorkis

When I asked my broker in January 2008 if we should go to cash and he said no—and I went with his answer. —@dldnh

Not having a better alibi. Last Thursday. —@yodaveg

Cutting off all communication with a guy after turning him down. Months later, plane he was flying crashed & he died. —@sleeplessinkl

Not learning Spanish—now it's too late (until eternity) to understand my grandmother's incredible and funny stories. —@robsaenz

Letting someone else influence my decision and missing out on a major life goal... —@Wrench06

Dropping out of school! —@jadawa

Supporting someone in her decision to leave her job, and having it go horribly wrong. —@Boxianna

Not having kids earlier. Had my 1st @ 39 after 5 yrs of infertility. If started sooner, maybe could've avoided infertility & had more kids. —@Doublelattemama

Turning down that offer of a free ride in a stunt open-cockpit biplane when I was 15. —@sailorgrrl05

Not taking my boyfriend seriously enough when he asked me if I thought we were soul mates. He died a year later. —@loveyourspirit

Buying into HD-DVD. Even worse, some of the movies I bought in HD…"King Kong," I'm talking to you. —@betaboy78

Not paying attention to more family stories growing up! —@scottgina

Not kissing Char when I had a chance that night. She liked me and I had no idea. I was 14. My love life would have changed forever. —@pixelshot

Voting for Bush in 2000 & 2004. I saw the light in 2008 (too late). —@mchessler1

Being shy in high school. Sooooo many missed potential date opportunities! —@AkiIskandar

Two regrets: (1) never took that gig as drummer for Genesis, and (2) never got offered that gig as drummer for Genesis. —@robertgdaniel

Thinking I'd have a second chance to give my big brother a real goodbye hug. —@MrsRoadshow

Not keeping a journal throughout my life. —@sassyscorpio71

Not keeping my money under my mattress. —@benjiejr

Keeping my Blockbuster membership. —@johnmarkharris

I regret that it took me so long to understand how important & exciting it is to take risks. Don't let fear hold you back! —@live2learn

Not applying sunscreen in the summer of 1978. —@grabbingtoast

Buying a controlling stake in the Planet AIG line of novelty restaurants. —@Notsewfast

Leaving a good friend without saying goodbye—19 years ago. —@halophoenix

Greatest regret will always be not listening more to my grand-father's brilliant stories before he passed away. —@bnl771

Waiting for a comfortable solution. —@DBtech

Never telling my college sweetheart what I felt. Now she's married with two daughters. —@Marcodj

Not visiting New York before 9/11. It was my dream to visit the Twin Towers... —@iNss

Not kissing that girl on the train journey home. (Forgive me for not being too specific.) —@XmasRights

I regret all the time I wasted regretting. Wise advice: You made the best decision you could with the information you had at the time. —@susanchamplin

Invent a formula for disaster.

bucket of water + cat toenail clippers + anxious feline —@squealingrat

life insurance salesman + stuck elevator + now empty super-sized diet pop – air conditioning —@A3HourTour

iPhone + jury duty – charger —@alaskanjackal

(mommy's white pants + toddler) / lunch —@MetaMommy

Hamlet – father + ghost + neuroses —@pumpkinshirt

extreme weather + land mass + world's poor – government assistance —@katoshiko

AIG + poor regulatory oversight + endless amounts of money – humility —@stewartwilner

midnight glass of water + open drapes – pants —@pumpkinshirt

2 teenage boys + loaded BB gun + trampoline —@roderickrussell

(Bush x 8) + (Cheney x 8) + (Rumsfeld x 6) —@windyhill

brand-new iPhone + 9-year-old + shirt pocket + cartwheel —@jakemates

(thunderstorm x loss of power) + fridge full of groceries – TiVo recordings —@hriefs robots + human emotions —@HayleyLovesYou

3-year-old + ice cream cone + crowded airport tram – smooth braking system —@VenetianBlond

subprime mortgages (homes) + mortgages (homes) + (30x leverage) – regulation – income – savings—@ophiesay

big SUV + small garage parking spot + sturdy concrete post – $300 deductible —@hriefs

car trip + relatives + talk radio – book

—@pumpkinshirt

unemployment + Twitter + Scotch —@kizel

14 six-year-olds + minigolf course + lots of sugar —@dances_w_vowels

2 cookies / 3 children —@Ttocs1817

nerves + red wine + significant other's parents – food —@dawns8

10-yr-old + scissors + glue + puppy – adult supervision —@plocke

Vicodin + 4 shots of vodka – common sense —@superAL1394

spaghetti + white shirt + gesticulation —@pumpkinshirt

ruby red lipstick + very white bridal gown + sudden limo stop —@singersorel

husband + toddler – common sense —@FrostKL

1 microphone / (Bill O'Reilly + Keith Olbermann) —@srujan

(ex-husband + 10-year-old x stubborness) / remote control —@plocke

coffee + stoplight + open laptop on passenger seat —@MOONLYTEN

"The Sopranos" – cursing – nudity + commercials —@pumpkinshirt

nervous sixth-grade boys + Human Growth & Development (aka pre–sex ed) + eggs for breakfast —@TiffanySchmidt

teenager + college – parents —@TomBrend

(Engineer + Marketing + Legal) x 3-hr meeting —@raccettura

mad brother-in-law + powerful laxative + long drive through eastern Wyoming —@Tykerman1

pantyhose + weight gain + class reunion – Spanx —@brucefreeman

electric fence + dog + metal leash —@MissXu

me + a girl—@pathumx

"Phish"... "troll"... Make up a NEW term for something that happens on the Internet.

Downloafing: Deciding you won't do any more work until the files you're downloading are complete. —@pumpkinshirt

BuzzButt: Getting Twitter updates constantly on your cell phone, which you keep in your back pocket. —@Sandydca

Rantique: An outdated comment posted long after the blog is written and the thread has died (or my tweet responses). —@jwenderoff

Googleplexy: The angry frustration of finding, by Googling, that something you thought you cleverly made up already exists. —@dudgeoh

Commental (adj.): Leaving an excessive number of angry comments on a blog post. "He's going all commental on Pogue today." —@noveldoctor

Post Mortem: A post telling everyone that someone has died. —@yodaveg

Echohoho: An e-mail joke that keeps getting forwarded. —@TonyNoland

Bullock!: An expression of joyful outrage over movies that really misunderstand technology. —@pumpkinshirt

Leftove: tweet that won't fit in no matter how much work you put into reediting i... —@yodaveg

Wikid: Something that sounds totally cool but is only 73 percent likely to be true. —@noveldoctor

Johnny-come-l33tly: Someone who discovers Internet slang a year after everyone else, but doesn't know it. —@pumpkinshirt

Search-based hypocohondria: The belief that you have the deadliest disease that fits with the symptom you've looked up online. —@yodaveg

Twicking: Quickly clicking on a work application to hide your Twitter page when your boss walks in. —@dirtybird1977

Tangential discombobulation, or TD: Hopping link to link until you forget what you were originally researching. —@pumpkinshirt

Para-site: A web site that derives its content by leeching from others. —@DSpinellis

Repalling: Replying to all in an email message. —@joanfeldman

Wikipedestrian: Crowd-edited to the point of bland banality. —@pumpkinshirt

Limbosphere: When it takes 20 minutes for an email to arrive, this is where it's been. —@yodaveg

Peacocking: Pulling your new MacBook Air laptop or iPhone out at a meeting or in the airport lounge to win admiration. —@yodaveg

Windows shopping: Filling up online shopping carts & closing your browser before buying. (Disclosure: I'm a Mac user! Still guilty.) —@_wendy_r_

Freetweeting: When someone retweets your tweet but doesn't give you credit (e.g., no "RT"). —@MarkRosch

What's the best advice your parents ever gave you?

It's just as easy to fall in love with somebody rich. —@craigwink

Life is not a dress rehearsal. Live it as if it's your only take. —@jcordeira

Quit early, quit often. (Infinitely useful for software development.) —@2dkid

If you can't be good, be careful. If you can't be careful, be good. —@wavedeva

Son, there's a time and a place for everything. It's called college. —@BarrSteve

There is no such thing as bad weather—just bad gear. —@justinide

In 5 minutes, you'll be 35. —@paulgreenberg

Never pass up an opportunity to go to the bathroom. —@_hillary

Don't let school get in the way of your education. —@Navesink

Even if you grow up to dig ditches, be the best ditch digger you can be! —@scottydc1

EVERYTHING is negotiable! —@strockman

My dad's response to many situations, from his Army Ranger training: "Reach in, pull out a handful of gut, and move on." —@thinkc

Do everything in life in moderation. Including moderation. —@bobvoisinet

Always shine the back of your shoes. It's the last thing they'll see as you walk away. —@PeterWeisz

If you caught a fish every time you went fishing, it'd be called catching. —@BruceTurkel

Always wear clean underwear. You never know when you'll end up in the ER. —@AndreAguiar

When I was a new driver, my dad told me: A ball rolling in the street is always followed by a child! —@louisekforman

Don't worry about the price of things; worry about the value. —@DrKoob

If you buy the dress, the occasion will arise. —@haejinshin

You are not the master of your fate. (It's good advice.) —@rosemheather

Mom on dating: Never order spaghetti, and always bring money for carfare home. —@honk4peace

The customer is not always right, but don't let him know that. —@Garyboncella

Marry a guy who treats his mother well. —@Leslie2823

Always keep $20 hidden in your wallet & your car. (Best advice my dad ever gave me. Saved me countless times!) —@LayersTV_RC

Nothing is free in this world. Everything will be paid for at some point. —@coolcatplayer

Don't despair. No matter how bad things get in life, it could always be worse—and it will always get better. —@dustin_j

Enjoy it, but don't get used to it. —@heimm

Always do the right thing...even when no one is watching. —@bobrall

First engage the brain, then the mouth. —@hush6

Never buy the first-year model of anything—car or doodad. —@James_Amos

Don't be concerned about self-confidence; that will come. Be concerned about self-respect. —@Scoobie

Wait one year before getting a tattoo. If you still want the same thing then, go for it. (Needless to say, I did not!) —@SadieBug

Don't be a jerk to tech support when you're late for a presentation. —@ctkennedy

Nobody gets through life unscathed. —@laurelhart

Separate the lights and darks when you do your laundry!
—@hobbypark

From my mother: Enjoy sex while you're young—just don't let it interfere with your schoolwork. (She was born in 1918!)
—@megsaint

Have fun every day. —@seancorcoran

Nice doesn't cost anything. —@chemrat

Separate bank accounts. Always. —@katerock7

If you don't let the cat get the string once in a while, the cat will stop playing with you... —@CarlTramontozzi

When in doubt, throw it out. (Now I live clutter-free.)
—@JoannaAven

Use the phrase, "You might be right" to end a disagreement. Works great in a marriage. It doesn't have the "Yes, dear" connotation. —@tweetkatie

Walk in the crosswalk. If you don't and you get hit, you can't collect insurance. —@TheGimpyGirls

Be nice to every girl now, because you never know who will end up the diamond from the bunch. —@MSGiro

Music can be a very tough business. —@Jeremy_Robinson

Date someone you can live with. Marry someone you can't live without. —@WickerParkGuy

It is better to keep your mouth shut and be thought a fool, than to open it and remove all doubt. —@McSteavenson

Never start a fight…but always finish one. —@genejm29

Why hang with the turkeys when you can soar with the eagles? (Took me 'til my 30s to understand.) —@alizasherman

If someone offers you a breath mint…take it. —@dsr

Pack your own chute. —@jillmcc

You have heroes all around you; choose the right ones to admire. —@modF

I expect you to screw up. Just don't screw up badly. —@rockivist

If you're not five minutes early, you're ten minutes late. —@Mugsie84

2 eyes, 2 ears, 1 mouth. Use them proportionally. —@pcz

You will never get rich working for someone else. —@skozeny

My dad was a cop. He always told me, "Never go back to the scene of the crime." —@smccormack

Never argue with a woman. You won't win. Even if you win, you lose. —@foolsprogress

If 5 people tell you that you're drunk, lie down. —@gzicherm

If at first you don't succeed, we'll still love you. —@zwb

What's your brilliant idea to improve air travel?

For every minute your flight is late taking off or landing, you get a $1 rebate. —@jeremyb

Allow passengers to strip completely naked for the duration of the flight. Breeze through security, and no need for in-flight movies! —@Notsewfast

e-tickets should be really electronic-tickets...no more printing! —@planetcho

Separate the cabin from the plane! No gates; board cabin in airport, slide it into plane. Comfortable, leisurely, quick plane turnaround! —@gielow

Stop telling us how to use a seat belt!

—@BouPierre

Aisle-long slip-and-slide. Roll out plastic sheeting. Add beer. —@rolando

Your frequent flyer card can be used to lock the seat in front of you in the full upright position. —@shayman

Place solar cells onto the wings & body—you could probably soak up some nice juice when above any cloud cover. Only day flights. —@McEuen

Fly-on-your-own-responsibility airlines that allow you to bring guns and liquids like in the good old days. —@hagnas

National Bank of Airmiles—buy, sell or lend miles, etc. —@passageC

Have the onboard PA system double as an in-flight karaoke! Imagine the fun we'd have! —@badmanj

Start disembarking from rear to front. Get up, get your bag, walk. That way you don't have wait for the slow people in row 10. —@desmodus

Have a better-than-first-class seat IN the cockpit. One or two will suffice, and cheaper than space tourism. Wouldn't that be cool? —@tecnocato

Board window first, then middles, then aisles. No more aisle clogs. —@hose311

Add a soundproof Kiddy Koach section. Fun for the kids! And a relief to the adults! —@bnl771

Open an airline that makes you sign a waiver stating that you are at peace and ready to die. Then allow for security-free boarding. —@Notsewfast

For people who can't afford Economy, create Steerage Class in luggage hold. Include parka and oxygen bottle. —@BarrSteve

In bad weather, when the seat belt sign is on and off, the flight attendants should hand out numbers for the bathroom queue! —@jeanniez

Real books instead of SkyMall catalogs. —@CBielstein

Footrests in coach! If even buses can have them, why not planes? (Been waiting for someone to ask me this for years!) —@AnnieLaG

Replace swim vests with mini-parachutes. Similarly useless, but, for most people, more reassuring. —@DSpinellis

Easily (electronically) reconfigurable distance between seats—variable legroom. Then charge by the quarter inch. —@oscartoro

Passengers with connecting flights should be always positioned close to the doors to improve speed in these cases. —@Alex_at_Itautec

Raise the seats an inch or 2 higher. Keeps the same distance from seat to seat, but gives you a bit more legroom. —@nocturne1

Offset the seats to provide an armrest solution. —@GlennF

In-flight Bingo. With swell prizes! —@noveldoctor

Special family-vacation and biz-travel days. Never the two should meet. —@IARSprint

Let non-ticketed people go to the gate again. Those wistful goodbyes and joyous greetings aren't as good at security checkpoints. —@CharlieLevenson

Snacks for the adults, sedatives for the children. —@justbustr

Boarding pass doubles as a key to your personal carry-on storage area (no one else can put their bags above your seat). —@daveabraham

Charge by the pound. Ends obesity epidemic, saves fuel, rakes in the dough in first class! —@McEuen

Use the inflatable slides for disembarking ALWAYS—faster and more fun! —@cvanheest

Instead of charging for checked bags, charge for carry-ons! Incentive to check, speed boarding/deplaning times. —@benmcallister

General anesthesia.—@megsaint

What made your first kiss memorable?

First kiss: 5 yrs old, at kindergarten naptime. I ended up with chicken pox. —@molliebush

His mother had to drive us to/from freshman high school dance. Kiss at the door—then his mother backed over my mailbox as they left. —@macnbc

In swimming pool, both have teeth braces, wave machine slams us together, braces lock, kiss happens. There was blood. —@talkinape

I said, "That made up for too long waiting." She thought I said, "We made out for too long, baby." She ran away. —@wordlesschorus

I was so nervous, I got dizzy as I leaned in to kiss her. What resulted was more of a head butt than a kiss. —@jadawa

7th grade, "7 minutes in heaven" game at party. He gently turned my head to kiss—my neck cracked like knuckles, one after other. Mortifying. —@MDTeresa

She lived far from my house, so I took the bus; I was so excited about it that I lost track of time and ended up in another state. —@iyttahm

The girl went on to become one of the stars of a "Real House-wives" reality show. —@Billforman

My 1st kiss, she forgot to put car in Park—slowly ran into my garage during kiss! Hard to explain to my parents.
—@betaboy78

First kiss in high school, in car, front of her house. Leaned on the horn as we kissed—yes, her parents were at home.
—@ChaseClark

It was in front of an auditorium of 500 people—including my parents! (school play) —@EvanFogel

Two first-timers. He missed, kissed my nose. Embarrassed. I said, "If we don't practice, we won't get better!" Tried again—fireworks! —@leleboo

Describe your 15 minutes of fame.

I juggled for Clinton's inauguration. 20 minutes of FBI pat-downs, and then I wound up throwing knives around the president anyway. —@McEuen

I'm on a Girl Scout cookie box (have been for 9 years, so it's longer than 15 minutes)! —@libbyfish

I was in an outhouse that was blown over by a plane taking off at an airshow. It was messy, but I got free tickets for the next 2 yrs. —@designerbrent

I was Miss Teen NH during an election year. I posed with every also-ran presidential candidate. —@trixielatour

I sang on Stevie Wonder's 1979 album "Journey thru the Secret Life of Plants" (Track 6—I'm one of the Japanese kids). —@takakon

David Letterman kissed me on air on his 6/17/93 NBC show; then I bumped Cindy Crawford off the show. Made the "Daily News" in NYC. —@GrumpyHerb

Andy Williams asked me to sew a button onto his shirt, then Glen Campbell sang a bit of "Galveston" to me over the Astro-dome speakers. —@cafepressmemaws

A friend wrote for "Seinfeld," used my name for a fictional company in an episode (great one, too). (Technically, 22 minutes of fame.) —@babydaze

I was Michael McKean's body double in an independent film. He was 60 years old and I was 24. Opportunity of a lifetime. —@danny_shea

My stepfather was "The agony of defeat" guy on "Wide World of Sports," before the ski jumper (he was the spinning-out Daytona 500 car). —@BigDaddy978

One of my iPhone web apps was a Staff Pick on Apple's website. I couldn't believe it! —@gizmosachin

I was on "Romper Room" (you have to be over 40 to know that one). —@BetsC

Lambasted by a right-wing radio host for teaching a course on the Klingon language. —@AaronBroadwell

Age 15, I played drum solo at Academy of Music in Philly. Got a standing O. Next day, I was fired from school band for missing the spring concert! —@brmperc

Volunteer fireman; national airtime rescuing ice fishermen off Lake Erie. Video of me walking a man off the ice. —@justcombs

I won the "American Way" magazine Road Warrior MVP contest in 2006. I was featured in the magazine, too! —@mordy

I was a runway model at the Playboy Club. I was 12. We modeled snowsuits. It was a 4-H fundraiser. —@haikumom

I appeared naked in a New York Times Sunday Travel Section photo, 6/26/94. (No names or obvious body parts appeared, though.) —@farinata

When my family & I were chosen to open Disney World for the day, I threw my confetti prematurely. Teenaged sons were mortified. —@mandomoose

Rescued drowning man while in Germany. Interviewed for TV and newspapers. Had to return home before they were aired. Never saw them. —@joshuaentwistle

I marched in Obama's Inaugural Parade with a contingent of returned Peace Corps volunteers. I carried the flag of Sudan. —@wlerik

I bought a house where someone was murdered. 1st year: endless drive-by photos, interviews, ABC-TV special... —@lsmith1964

I sang backup and played tambourine on stage with RuPaul in the late '80s. STARBOOTY! —@cswanger

I was crowned Miss Watermelon at local pageant as a teen. Parade, interviews...embarrassment to last a lifetime. —@sgoodin

What's the weirdest job you ever had?

Selling accordion lessons door to door in Lincoln, Nebraska. —@esklf

I'm a sword swallower. —@roderickrussell

Aquarium gravel packer. —@lukebrindley

Answering phones at a brothel. Just a few hours (no, not my line of business, nor a customer; long creepy story). —@victoryfarm

Once during a summer internship, I was Buddy, the Tyson Foods chicken. —@Fletch_80

As a teen, I watched TV for Gallo Winery. I monitored commercials that aired before and after Gallo's. —@cynthiamckenna

Age 19: Mall Easter Bunny at photo booth; helping circus advertise by riding elephant around mall in rabbit suit. —@A3HourTour

I worked in a pickle factory as a teen. It was hot & rank. I can't enjoy pickles today. —@passepartout

Bowling-ball hole driller. No mistakes allowed. —@stevegarfield

Department-store bathrobe model at Christmastime—loved the looks I got riding the escalator in pink robe & fuzzy slippers. —@debbieduncan

Lighting designer for pornographic magazine photographer. Boy, THAT was a train wreck! —@IowaShorts

Slug-washer for a neuroscience researcher (you know, like garden slugs). —@shrinkraproy

I had to run out and clean any spots off the Wheel of Fortune during commercial breaks. —@carolpascale

I worked one summer as an intern at the Liberace Museum. That experience can't be summarized in 140 characters. —@danmcq13

Proofreader for a skywriting company. —@shiveringgoat

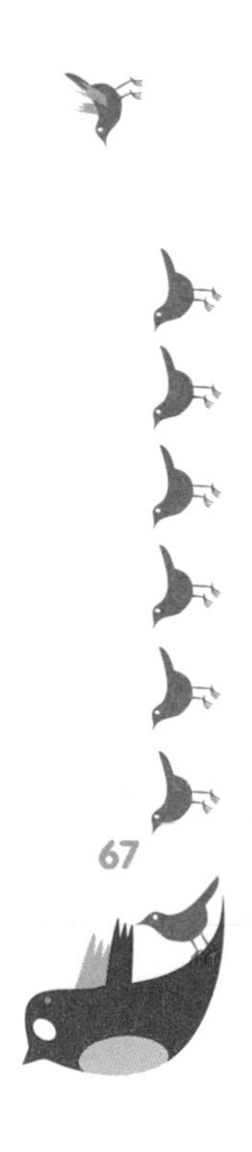

Balling machine operator, during high school. I ran the machines that attach ball ends to Ernie Ball guitar strings. —@mikemaughmer

I used to freeze dead people for a living. —@roderickrussell

I once had a job delivering drill bits to dentists' offices on a scooter. I know nothing about dentistry (or operating scooters). —@unmiked

Horse artificial-inseminator (when I was in college I studied equine reproduction). —@kari_marie

Getting pigs, cows, and sheep to pose for photographs. —@hannnahkl

This HUGE company paid me to walk around the building at 2 a.m. and flush toilets. YUP. —@EllaJo08

I had to put empty toothpaste tubes on an assembly line to be filled… —@MichaelRubin

Driving Snow White and Prince Charming, atop motorized wedding cake. On ice. —@reganref

I was a hooker…I bent the candy canes into perfect hooks at Logan's Candies. —@mmmwaffles

Make up an original Tom Swifty. (Example: "There's the lamb's mother," Tom said sheepishly.)

"I wish I remember being at Woodstock," Tom said acidly.
—@WConnolly

"This bank is too big to fail," said Tom balefully. —@Mainesailor

"Watch out for the poison ivy," Tom said rashly. —@jschechner

"I dropped the cashbox overboard," said Tom sanctimoniously.
@garrymargolis

"I hope this whole state falls into the ocean," Tom said callously.
—@MMmusing

"I remember when I almost won that election," Norm Coleman recounted. —@stevenaverett

"I didn't get confirmed as Supreme Court justice," Tom said disappointedly. —@cstorms

"This weight-loss diet has really worked!" she expounded. "It didn't work for me," he announced. —@mtobis

"This is the perfect pen to grade papers with," said Tom incredulously. —@pumpkinshirt

Make up an original Tom Swifty. (Example: "There's the lamb's mother," Tom said sheepishly.)

"My new MacBook is as light as a feather," Tom said airily. —@ericd1

"America Online won't let me send an instant message no matter what key I press," said Tom aimlessly. —@pumpkinshirt

"Hey! No cheating at Tic-Tac-Toe!" Tom said crossly. —@mehughes124

"I've just dropped all the paperwork off at the insurance office," Tom exclaimed. —@cvanheest

"I'm opposed to all camping equipment," Tom said contentedly. —@pumpkinshirt

"Stand and Deliver is a much better song than Goody Two Shoes," said Tom, adamantly. —@oneill_colin

"Here's the pitch," he said underhandedly. —@SkyDawg1971

"Why weren't 'The Hulk' and 'Brokeback Mountain' as good as 'Crouching Tiger, Hidden Dragon'?" Tom asked angrily. —@JaredParker

"Soft woods rule!" Tom opined. —@pumpkinshirt

"My mother's German," Tom muttered. —@porousborders

"My horse done run me over," Tom said, downtrodden. —@jschechner

"Use the service entrance," Tom said indecidedly. —@hughesviews

"Whatever happened to Terpsichore?" Tom mused.
—@susandetwiler

"The patient is in cardiac arrest!" Dr. Tom said heartily.
—@jschechner

"Who left these shoes on the staircase?" asked Tom trippingly.
—@marlaerwin

"Another toll booth?" Tom asked feebly. —@jschechner

"Time for my flu shot!" Tom said innocuously. —@jschechner

"I used to have a can-do attitude," Tom said candidly.
—@ColinDabritz

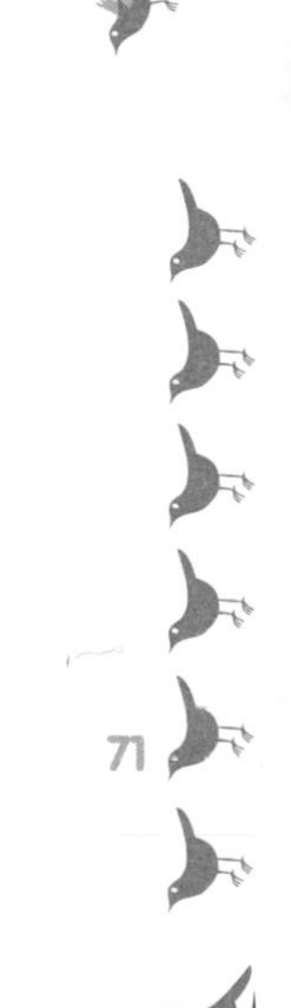

"I'm not going to school today," Tom said absently.
—@OMPNewsEditor

"Is Goldie Hawn still married?" Tom asked curtly.
—@robertgdaniel

"I can tell you the gender of any insect in that mound," Tom said buoyantly. —@pumpkinshirt

"Where is Darfur?" asked Tom suddenly. —@Larrco

"I'm the new goalie on my hockey team," Tom said puckishly.
—@mmcd908

"I do so love Roman aqueducts!" Tom said archly. —@d_mcmullen

"This round's in honor of the 'Saturday Night Live' executives," Tom said forlornly. —@pumpkinshirt

"Who took my trousers?" Tom panted. —@_Bone

"Do you mind if I take your bread, man?" Tom asked gingerly. —@d_mcmullen

"These 140-character messages are quite something!" said Tom, all atwitter.

—@michaelmarcus

"Diabetes is a serious matter," Tom replied insolently. —@Notsewfast

"You'll grade that exam twice!" Tom remarked. —@Notsewfast

"I can't put the stereo back together," Tom said inconsolably. —@hughesviews

"Thinking of purchasing a German car," Tom declared audibly. —@jappio

"My tree's branches have sprouted again!" cried Tom, relieved. —@michaelmarcus

"I work for a chip manufacturer," said Tom intelligently. —@bpdobson

"My keyboard won't type capital letters," Tom said shiftlessly. —@kskwriter

"The fire was so hot it even destroyed my cookware," said Tom with panache. —@pumpkinshirt

"I'm gonna build a church when I grow up!" Tom said aspiringly. —@bpdobson

"Throw it off the back of the boat!" Tom said sternly. —@RobTuck

"They've made me trade roles with my understudy," said Tom, downcast. —@miche

"You CAN stop being a prostitute," Tom exhorted. —@yodaveg

"I say who falls in love," Cupid said arrogantly. —@MMmusing

"The ambulance is here. Stand back and give him room to breathe," Tom said summarily. —@hughesviews

"I'm not ready for the digital TV conversion," Tom said blankly. —@MarkRosch

"I'm going to memorize the Gettysburg Address," Tom said ably. —@MMmusing

"That's so meta," Tom said swiftly. —@JaredParker

What's the cure for hiccups?

Put a cold spoon on your back—that's what my grandfather would do for hiccups. —@florian

Put your head between your knees and swallow hard. —@megs_pvd

Drop a lit match in a glass of water to extinguish it. Take out match. Drink water. —@michaeljoel

Have someone slowly & softly count backwards from 10 to 1 in Russian for you. Works every time! —@rgalloway

Check your 401k. That should scare the hiccups right out of ya! —@warcand

The cure for hiccups is simply to get the air out of your stomach. How is up to you. —@drct

Take a glass of water, hold your breath and gulp it down. Distraction helps against hiccups. —@kashaziz

There's gotta be something in the App Store for it by now. —@aaaaiiiieeee

Sounds crazy, but it works. Take 9 sips of water then say, "January." Laugh now, but you'll thank me when the hiccups are gone. —@garmstrong65

Peanut butter on a spoon. —@erlingmork

Plug your ears and nose and drink seven gulps of water. Difficult, but doable. Works like a charm EVERY time. —@amysprite

With right hand, reach around to behind left shoulder tightly and grab some back flesh, hold for up to a minute and no hiccups. —@SullivanHome

Promise yourself something you really, really want (and mean it) if you do hiccup again. It works! —@jillgee

Hold your breath and go slowly thru ABCs. Then at Z, take another deep breath without exhaling. Then slowly exhale. —@assignmentdesk1

Drink out of the far side of a water glass (best done over the sink). Works every time. —@DavidWms

Dry-swallow a spoon of granulated sugar. The trick is to overwhelm the overstimulated vagus nerve (causing hiccups) with new input. —@enrevanche

Eat a full spoon of crushed ice. —@JuanluR

Rub both of your earlobes at the same time. Hiccups will go away. —@tiffanyanderson

Try drinking a cup of water with a paper napkin over it. I swear it works! —@SocialMediaSabs

I take large sips of bourbon. It doesn't stop the hiccups, but I stop caring! —@Chiron1

Describe your Most. Embarrassing. Moment. Ever.

Co-worker asked me if I had a sec to help a customer. My reply: "I have lots of secs!" Yeah. MOST. EMBARRASSING. MOMENT. EVER. —@sassyscorpio71

I fell through ceiling into the bathroom, taking the sink and toilet out, seconds before boss said, "Be careful, there are loose boards." —@ACJohnR

I once interviewed Ed Bradley. Asked, "What's the most interesting place you've been?" Ed: "Khyber Pass." Me: "The bar in Philly?" —@Jimbrez

1st date (with now wife). I drink from straw, don't want to break eye contact, straw goes straight up my nose & I bleed like stuck pig. —@UNnouncer

As a church musician, I tried to swat annoying fly as it landed on female singer. Got handful of boob instead, as congregation watched! —@hughesviews

First week of my new job, I accidentally signed off an email to my male boss: "With love, Jess." (And yes, it was a total mistake!) —@jessica_allison

Sent letter to our biggest client, included heading: "Your account" (secretary left out the O). Tricky meeting followed. —@ctrly

I worked info desk at mall in Stamford. Family asked for info and I said, "I have their card right here"—and pulled out a condom! —@ascottfalk

I said, "I hope she does not leave" about seriously ill coworker who might have to quit her job. But my accent sounded like "does not live." —@girl_out_there

My HS swapped boys/girls locker rooms over summer. I didn't know, showered in the wrong one. Empty when I got in, not when I got out. —@danfrakes

The fly on my suit broke before a daylong interview with a female boss at HP... interviews have been easy ever since. —@chris2x

Library user asked for an 8x10 color photograph of a kabuki. I thought he said a "wookiee," and proudly gave him a photo of Chewbacca. —@carolcdt

On date, I fell into a waiter w/full tray. Read about it later in "Glamour" when guy wrote about his most embarrassing date ever. —@SusanEJacobsen

Friend e-mailed about first job (sent to family, friends, grand-parents). I accidentally hit Reply to All, writing, "About f*^^%ing time!" —@alechosterman

Lesson learned: Before sending a press release out over the wire, make sure that "public relations" is spelled as intended. —@hriefs

In 2nd grade, I was skiing in Utah and wet my pants! But the best part is it froze on the outside!! —@yoshionthego

While arranging a date, I accidentally left out the 2nd word of "What time do you expect to get here?" —@laroquod

I underestimated my own hair and bought only 1 box of dye instead of 2. I had a fox tail instead of a ponytail for a week. —@gnarlykitty

Spent $$$ on Saks business suit for 1st meeting at United Technologies. The back ripped while I was at whiteboard presenting. —@IdaApps

Age 17, I was Santa at Kmart. Me: "What do you want for Xmas, little boy?" His mom: "He's a SHE!!" (Looked like a he!) —@choirguy_

Waiting for a crescendo in high school band practice to break wind under cover of music. Conductor said, "Cut!" I did. —@pmahoney87

I fell into the fake river at a mini-golf course once. Came out blue from the coloring they put in there. Embarrassing enough? —@pithe

'64 Buick V8. Accelerator stuck, smoke billowing. Run to gas station for help, run back, mechanic follows, yells, "Turn off the KEY!" —@sparker9

Discovering AFTER interviewing Billy Crystal that my skirt had been unzipped the entire time. Not MAHvelous.
—@susanchamplin

New car, moonroof out in the trunk, decide to install it while wife driving, moonroof flips away, smashes into 1m pieces.
—@Targuman

Instead of "It's too important for that," I once typed "I'm too important" & sent it to 4 people. —@laroquod

I once farted at an autopsy, with a dozen medical students in attendance. —@seaslug_of_doom

Explain a facet of modern life in the style of Dr. Seuss.

A call to my bank
So simple I think
Can't get to a human
Now I need a shrink! —@AndyMaslin

A netbook, a MacBook
An 8-gig iPhone
A trio of toys
Needs a singular loan —@superbalanced

One fish, two fish
Red fish, genetically engineered grain-tolerant blue fish
Patent Pending, Globo-Fish Systems, Inc. —@pumpkinshirt

The markets went through
Some precipitous drops
'Cause we put all our money
In credit default swaps!
—@keylager

TV, yes TV
For you and me
Hundreds of channels
And nothing to see —@vdr

Would you like some, my good man
Would you like some processed ham?
You will like it, you will see,
When we drink some processed tea! —@AndyMaslin

Foodies very loud and vocal
Scream if it's not homemade and local! —@elizabeth627

To get attention
Limbaugh rants
And Paris wears
No underpants —@VenetianBlond

Pogue Sez

I love my Twitter followers, I really do. I doubt that you'd be able to find such an enormous group of bright, witty, articulate people in one "place" anywhere else.

I was surprised, however, to see how much trouble most of them seemed to have with poetic rhythm. The notion of consistent meter ("da-DUM da-DUM," etc.) seems to have escaped most contributors. (Remember, the entries printed here were the *best* attempts.)

Clearly, it's a rare gift. Apparently, Dr. Seuss was Dr. Seuss for a reason.

I saw him, I saw her
I saw it and saw you
And to think that I saw it
On Google Street View!
—@pumpkinshirt

One bar, two bars
Three bars, four
My iPhone's signal
Requires more
Yes, sir, yes, sir
I do mind
I've no reception
Of any kind —@jsnell

MSNBC's lefty
Fox News to the right
When they want big ratings
They fight, fight, fight, fight,
fight! —@dhersam

Just bought Thing 1
It's shiny and new
But soon obsolete
With release of Thing 2 —@UncJonny

I've dropped you from Facebook
MySpace, all things 'Net
We're no longer friends
(Although we'd never met)! —@valsadie

Damn, I am
Descartes, not Sam. —@LibraryBarbara

I mail, I text, I tweet, I blog,
I build a Facebook for my dog,
I speak no words, I shake no hands,
I am at last a modern man. —@smacbuck

"LOL" and "BRB" are decades old. Make up a NEW Internet abbreviation.

LOTI: Laughing on the inside. —@fatsvernon

TYL: Tweet you later. —@Vanish

BRB-JWCWT: Be right back—Just wrecked car while texting. —@brianwolven

WWYT: What were you thinking? —@SheilaBird

STWARP: Sorry, talking with actual real people! —@Remi_T

SAWTIA: Stop already with the Internet abbreviations. —@garygoodenough

ITDEC: In this difficult economic climate... —@DavidBThomas

PAFTA: Parents are following this account. —@mjestes

DWT: Driving while typing. —@dirtybird1977

MTTN: Milk thru the nose...for when something's funnier than LOL. —@BruceTurkel

RLF: Real Life Friend. —@dldnh

YWKHSLIC: You wouldn't know her, she lives in Canada. —@pumpkinshirt

QQJH: Quit quoting John Hodgman! —@pumpkinshirt

YAWAIARATIW: You are wrong and I am right and this is why. —@asdobkin

TCBTYBTCBY: This chat brought to you by TCBY.
—@pumpkinshirt

23S: Twenty-three skidoo! (Elderly chat-room participants only, please.) —@pumpkinshirt

qeafscxtvbfbyhninm: Cat walked across the keyboard. —@tuttleimages

WROFLBFIF: Would roll on floor laughing, but floor is filthy. —@pumpkinshirt

DYKHOIA: Do You Know How Old I Am? —@Sandydca

CJKI: Coffee just kicked in. —@pumpkinshirt

LITABBY: Look! Is that a bear behind you? (This trick only works on the truly gullible people online, not on the other 10%.) —@pumpkinshirt

DWATAT: Didn't we already talk about this? —@webenglishteach

MOMS: Moment of Mac smugness. —@pumpkinshirt

AARS: As Al Roker says... —@pumpkinshirt

BRBUSGIOT: Be right back, unless something good is on television. —@pumpkinshirt

Pogue Sez

In the end, several thousand people submitted entries to the questions in this book. But one stood out as a superstar submitter, both in quantity and quality.

I refer, of course, to @Pumpkinshirt.

He submitted 462 responses, all of them hilarious. After a brutal elimination process, we included 155 of them in the book. That's why so many of the tweets in this book—including so many of the "new Internet abbreviations" on these pages—seem to come from one follower.

Who is this guy? I don't really know. His Twitter profile says that his real name is Lain Hughes, that he lives in Mississippi, and that he's a "professional dilettante. Well, semi-pro."

What's he look like? How old is he? What's his writing background? What's "Pumpkinshirt" supposed to mean anyway?

Who knows? That's the beauty of Twitter: it's an absolute meritocracy. All we know about you is what you say—and funny wins.

In Pumpkinshirt's case, it became clear early on that we had a prodigious comic talent on our hands. If this book is the "American Idol" of one-sentence humor, Pumpkinshirt is the Season 1 winner.

PIWFJRCDS: Person I was following just revealed creepy dark side. —@hush6

GIIBSI: Great idea! I'll be stealing it. —@pumpkinshirt

HAY: How are you? —@afairfield

TL;DR: Too long; didn't read. —@DCTenor1

DTFMMLF: Does this font make me look fat? —@pumpkinshirt

TSRFCILBCISTROTF: This special room for chatting is lovely, but can I see the rest of thy farm? (Note: Amish chat rooms only.) —@pumpkinshirt

HIDOTA: Hope I didn't offend the Amish. —@pumpkinshirt

COMPTOW: Check out my poorly thought-out website! —@pumpkinshirt

SUNS: Shut up now. Seriously. —@ForeverMac

TUI: Tweeting under the influence. —@darwinexception

For Facebook: IDLHIJHS,E: I didn't like her in Jr. High School, either.
—@BruceTurkel

IATPOTKYTYWTT: I'm actually the parent of the kid you thought you were talking to. —@sheldonc

IRST: I'll regret saying this. —@eroston

NRPA: Not really paying attention. —@DerekNP8

TFBINALOL: That's funny, but I'm not actually laughing out loud. —@dzumini

ISGTT: I'm so gonna twit this! —@opica

NWF: Not worth finishing. (Good for technical reviews.)
—@bitfiddler

JKATHPI: Just kidding about that horrible personal insult!
—@pumpkinshirt

SFIGP: Sorry for the incorrect grammar and punctuation.
—@HarrisonWilson

TCABG: They can't all be gems. —@pumpkinshirt

What's the best toast you ever heard?

As you slide down the banister of life, may all the splinters be facing the right direction. —@mmarion

May the worst day of your future be better than the best day of your past. —@kriskelley63

A toast to karma and serendipity: May they have my cell number. —@Wylieknowords

May you never want as long as you live, and live as long as you want. —@SurHorse

To women, for without them we would have no reason to drink. —@bahamat

May your realities be even better than your dreams. —@etreglia

(Among a roomful of entomologists:) To the bugs...and their lovers! —@eBrenda

Champagne for my real friends, and real pain for my sham friends. —@NetEditrix

Here's to sobriety! —@acinonnap

Now thank we all the Lord divine,
Who turned the water into wine,
Look kindly on us foolish men,
About to turn it back again. —@vonschlapper

Here's to the man who takes a wife, for it makes a lot of difference whose wife it is he takes! —@WebWise1

I'd rather be here with you people than with the finest people on earth. —@erikzen

To our wives and our girlfriends...may they never meet. —@pshag

Redefine an existing word in a punny way.

Gyroscope: Ancient Greek instrument used to detect one's proximity to meat. —@sfreedman

Decagon: What happens to a seafront pier in a hurricane. —@DaveHyam

Blandishment: The word used to describe my cooking. —@livnvicariously

Benign: What you be, after you be eight. —@Imagin_that

Specimen: Italian astronaut. —@ahanin

Propaganda: To hold up a goose
—@colin_miller

European: Why you feel so sophisticated in the bathroom. —@donnaaparis

Morass: The greatest idiot you'll ever meet. —@nicolaguidi

Paradigm: $0.20. —@johnmarkharris

Elfin: A Mexican shark. —@justinfinnegan

Algorithm: An invitation from Al Gore's wife to play bass on Guitar Hero. —@JimFl

Cantaloupe: When the ladder doesn't reach.
—@smccormack

Consequence: Prisoners lined up in numerical order.
—@sboulton

Contradiction: Speech patterns of Nicaraguan counterrevolutionaries.
—@robertgdaniel

Discourse: In the style of a golf course, but where you throw frisbees around to get them into baskets instead.
—@jlawrencem

Discovery: The CD or DVD has ended. —@hush6

Duodenum: A pair of jeans.
—@jadawa

Elucidate: eHarmony match with someone who can actually speak in complete sentences. —@noveldoctor

Pogue Sez

If I learned anything from my Twitter-book experiment, it's this: Twitterers love puns.

My theory: Twitterers are funny in general, but the 140-character limit puts something of a damper on longer-form humor.

One night, I twittered: "TONIGHT'S MEDITATION: A cement truck has collided with a prison van. Police are in pursuit of several hardened criminals."

I should have guessed that wouldn't be the end of it; the follow-up puns turned the thing into a veritable crime story.

At first, things weren't looking good. Although "the criminals' mug shots are plastered all over town" (@gregoryahill), "there are still no solid leads" (@SamFK). "It could be a hard case to crack" (@natedelaney); "police are looking for concrete evidence" (@lhroadkill). "The investigation has hit a wall" (@imalindatoo).

Eventually, though, the case turned around. "The evidence against them is rock solid!" (@marchdigital).

"The charges are not yet set in stone" (@nburmandesign), but one thing was for sure: "They'll all be doing hard time" (@tonyfalcone).

Euthanasia: Japanese after-school programs. —@AbVegan

Fortnight: Old English castle guard. —@NJBrit

Intercourse: An onlinecorrespondence class. —@RosanneLambert

Heretofore: The current location of Jack Bauer. —@zigziggityzoo

Precursor: A Palm smartphone owner's finger. —@hughesviews

Interface: Winning an argument against your wife. —@MisterRoo

Onomatopoeia: When a dog relieves itself on the welcome mat waiting to go out. —@mcdermittd

Pokémon: To poke a Jamaican friend on Facebook. —@kylegray1

Universe: A song where the same lyrics are sung over and over. —@cmumathwhiz

Motif: What boxer Leon Spinks wishes for at Christmas. —@tomtwine

Taxidermy: Extra fee on plastic-surgery charges. —@hornsolo

Olfactory: A place where acronyms like LOL or ROTFL are made. —@livnvicariously

Recipe contest! You have 140 characters. Go!

Pan-fry 6 chicken breasts in juice from jar of marinated artichoke hearts. Add chopped 'chokes & 8 oz crumbled feta. Serve on orzo. YUM! —@wendallkendall

Pancakes by Ones: 1 cup flour, 1 cup milk, 1 egg, 1 T sugar, 1 tsp baking powder. Mix gently. Pour on griddle. Flip when bubbly. —@Tajmari

Pancakes for 60: 10 lbs flour, 2C baking pwdr, 3.75C sugar, 2.5 doz eggs, 1C oil, 1 gal milk. Combine dry then wet, mix, stir, cook, eat. —@selgart

1lb Velveeta + 1lb ground beef + 8oz taco sauce + 1 can Hormel Chili (with or without beans) = delicious, cheap, and quick dip! —@jshbckr

Pasta Aglio e Olio: Sauté 2 crushed/minced garlic heads in olive oil with one crushed red pepper. Spoon over pasta. Add locatelli romano. —@csunwebmaster

Husband Burgers: Wife prepares patties, tomatoes, onions & lettuce. He turns on gas grill, chars burgers & takes credit. She cleans. —@lizardrebel

Stir-Fry: Grab wok. Sauté garlic/onions. Cook meat. Toss in broccoli, water chestnuts carrots, etc. Add oyster sauce. Cook and stir. —@rissriss

Savory Grits: Sauté minced garlic in pan, + 1 cup H2O, + 1 tsp bacon bits (real). When boiling, +.25 cup grits. Boil til thick, + cheese. —@tsmyther

Unsloppy Joes: 1 lb ground beef, 1 can chicken gumbo soup, 1 T ketchup, 1 T mustard. Brown beef. Add rest. Simmer. Eat on bun. —@bonnyface

2 tbl sardines in oil/med-high heat, add garlic & 1 cup cherry tomatoes, heat until saucy and serve over spaghetti. —@kalunlee

Salmon Filet: Sprinkle sea salt, paprika, chili powder, oregano. Broil/grill on water-soaked cedar plank for 12–15 min. —@jschechner

Buttermilk Slaw: Toss 1lb. sliced veggies with whisked mixture of 1/2 cup buttermilk, 3T mayo, 2T apple cider vinegar, 1T honey, s+p. —@mizmaggieb

Partially Grill Grilled Cheese Sandwich. Cut round hole in middle. Drop in egg. Continue grilling until egg cooks. —@zvisus

Green Tea Vodka Spritzer: Highball glass of ice. Add juice of ½ lime. 1 oz vodka. Fill glass with Canada Dry green tea ginger ale. —@Gigipolitan

Marinate chicken strips in 1 cup yogurt + juice of 1 lime + 4 crushed garlic cloves. Sauté in butter with tarragon+szechuan pepper to taste. —@StMarksSTL

1 oz each gin, rum, vodka, tequila, triple sec, blue curaçao, 1 tsp sugar, juice of 1 lime. Shake well. —@schroncd

Fried Rice: Cook rice. Chop scallions. Combine all with 2 eggs in wok and fry. Add ketchup. (Yum!) —@Merts0812

Chocolate Mousse: Melt 200g chocolate. Cool. Add 6 egg yolks. Fold in 7 egg whites beaten to stiff peaks. Cool 2–4 hrs. Nothing else needed. —@girl_out_there

Uni Student Risk Meal: Take everything edible from fridge. Stew on low heat in a pan for many hours. Serve and enjoy! —@c_cooper88

Sauté 4T olive oil, 3 cloves minced garlic, 3pts 1/2'd cherry tom. Add S&P, oregano, basil. Cook 5 min, add cooked pasta & fresh mozz cubes. —@etreglia

Pan fry petrale sole (floured, salt, pepper). Serve over rice + snap peas (boiled 2 mins). Top with sliced almonds sautéed in butter. —@cdm57_2000

3 cups fresh basil, 4 cloves garlic, 1 cup olive oil, 1 cup grated parm, 1/2 cup pine nuts. All in blender, mix till smooth. Pesto!! —@jeanniez

Bacon Date Bombs: Wrap dates in bacon. Pack snugly on cookie sheet, seam down. Roast @400 'til well-browned, ~20 mins. Turn halfway. —@Recipitos

1 chicken, 2 onions, 3 parsnips, 4 carrots, 5 quarts water, salt, pepper, lemon juice. Boil then simmer 6 hours. Cool, skim, heat, serve. —@WFMTweetie

Best-Ever Whiskey Sours: 1 cup Jack Daniels, 1/2 cup RealLemon lemon juice, 2 tbs granulated Splenda. Shake over ice. Garnish with cherry. —@davidjan47

1 lb stew beef, 1/2 cup julienned ginger. Add water to cover both, and cook over med-high heat for 1 hr. Serve hot over rice or pasta. —@dwynath

Preheat oven to 375°F. Rub rib roast with thyme, rosemary, salt, pepper. Roast for 15 mins/lb. NO PEEKING! Rest for 20, carve and serve! —@dengelina

Homegrown tomato & lettuce, homemade mayo & bread, crisp toast & bacon. Assemble, squash together (Mom's touch). Summer food—BLT! —@Maggie_Dwyer

Pancetta-Wrapped Garlic: Simmer cloves 10 min or til tender. Drain, wrap in pancetta, secure with pick. Bake 12 min @400, til just brown. —@Recipitos

1 jigger Absolut Citron & 1 jigger Frangelico. Take as shot with lemon wedge dipped in sugar. Breathe out and taste the chocolate cake. —@PhotoBugF4

Husband + $$ + reservations = My favorite dinner recipe! —@cstorms

Summer Salad: 1C corn niblets, 1 diced avocado, 10 cherry tomatoes halved, 1 lime's juice, little olive oil, black pepper, and mix. —@citygal1

Bacon + Anything = Awesome. —@raccettura

Homemade Salsa: Combine 3 med. tomatoes, 1 small red onion (both diced) with 1 tsp. minced garlic, 1 tbsp. salt and 1 tbsp. olive oil. —@Katie_Traut

Cheese Crackers: .5C butter, 2C cheddar cheese, 1.5C flour, 1t salt. Mix all, roll out thin, cut into squares, bake at 325 for 11 min. —@eatingcleveland

Perfect Manhattan: 3 parts rye, 1 part sweet vermouth, 1 part dry vermouth 3 dashes bitters; Shake with ice. —@hadaly

Tarragon Corn on Cob: Wash 4 ears corn. Boil 10 min. Wrap in foil with butter & tarragon sprigs. Place on hot grill for 5–10 mins. Voilà! —@SusanEJacobsen

Mother's Little Helper: Equal parts vodka, triple sec, lime juice. Stir into pitcher of ice. Drink. Sigh heavily. Relax. —@smccormack

Sauté sliced bananas in butter over low heat till golden brown on both sides. Put on vanilla ice cream. Yum. —@billshander

Cobbler: Combine 1C each flour, sugar, milk. Pour in 13x9 pan atop 1 stick melted butter. Top with dollops pie filling. Bake 350 til golden. —@Sandydca

(1) Car, (1) Whole Foods, (5) Mile Commute. Drive car to Whole Foods' prepared food section, make selection, drive home = Dinner. —@BryonThornburgh

1.5C OJ, 0.5C vanilla yogurt, 1 banana, 1C frozen mango, 0.5C frozen blueberries. Blend. Yummy smoothie! —@virtuallori

Pour chocolate on it. —@balbes

What's your brilliant idea to improve the modern automobile?

Built-in breathalyzer. —@CathleenRitt

A sticker on the windshield or tag with your score on the driver's test. In huge numbers. —@KerriCW

DRIVE: Driver Reliability Intel Variable Equip. (Biometric wheel ID; system measures driver skill. Performance output based on skill level.) —@BerryLowman

How about a direct correspondence between the pedal and the acceleration? Cars that move with your foot off the gas freak me out. —@ZevEisenberg

Editable or downloadable horntones. So I can have Gandalf booming, "You Shall Not Pass!" when someone tries to cut me off. —@GoodOlClint

When you approach an oncoming car, a beeper notifies the other driver if he has his brights on. —@AndyMaslin

Backseat built-in wet-dry vacs...for all the food, drink, whatever my children spill in the car. —@MaryHr

High beams in the REAR of your car, for when you have someone behind you with high beams on, blinding you in your mirrors. —@justcombs

Heated wipers to prevent ice buildup on wiper blades during winter. —@AndyMaslin

Front-mounted brake lights, so you can tell if the idiot approaching you is actually going to stop. —@mklprc

License plate should show driver's cellphone number, so we can tell them to stop going slow in the left lane. —@politicsoffear

Car's paint color should change (à la mood ring) to indicate temper of the driver. —@cgranier

Built-in front-facing dash cam works like a TiVo. Records the last hour or so, to reference after accidents, report litterbugs, etc. —@applescriptguru

Drink holders that keep cold drinks cold and hot drinks hot. —@AndyMaslin

Design one with a place to put my purse. Never seen one yet. —@bibliobess

Put a limo cabin-partition window in all family cars. Parents need some peace now and then. —@NervousRobot

Eyeball monitor that sounds alarm when driver looks away from road (texting, looking through CDs) or closes eyes (sleeping, drunk). —@McCann_NY

Beacon that glows on roof if driver is texting or talking on a cell phone. Alerts nearby drivers. —@bgiese

Long-range red-light predictor: Tells you in advance how to adjust your speed to hit the next light when it's green. —@spannertech

Side mirrors that eliminate the blind spots. No more looking over your shoulder to change lanes! —@Dahveed7

A device that allows people to vote bad drivers off the highway, "Survivor"-style. —@Ophelia673

Digital displays on roof of car showing the speed of the car. —@normantowler

In mommy minivans: Integrated central vac system and refrigerated glove compartment. —@rachelbook

We need STANDARDS. Bumpers should all match up, car doors should all strike a bumper strip on the other car when the door is opened. —@azcurt

Headlights that automatically turn on/off/adjust based on ambient lighting conditions. Hey, MacBook keyboards can do it! —@flpatriot

Keys out, lights out. No more dead batteries. —@leereamsnyder

Paintball gun to mark offensive cars & warn others. Color coded—red: runs lights, orange: tailgates, blue: driver needs brain. —@Maggie_Dwyer

A pressurized air nozzle, powered by the engine, so you can inflate your tires without going to the gas station. —@aleidy

Variable-brightness LED brake lights—get brighter the harder you step on the brake pedal. —@iNss

Batman logo shoots up from the car at the push of a button—so you can find it in a parking lot. —@EvanFogel

Eliminate makeup mirrors. —@netback

Different horn sounds. Gentle for "Hey, did you see the light change?" Firm for "Whoa, watch it, pal!" —@bewildia

Cars would produce a mist, so you know who's honking. The fluid for horn mist would cost $7,000 a gallon. —@joshforman

Built-in message panels to tell that jerk EXACTLY how bad a driver he is.

—@hornsolo

Permit others to re-park cars that are badly parked. —@JRadimus

Less KITT, more Chitty Chitty Bang Bang. —@pumpkinshirt

The best way to improve cars is to improve their coffee-drinking, McMuffin-eating, eyeliner-drawing drivers. Drive the car better! —@HJBosch21live

One word: Carpoons. —@pamheld

Invent a witty collective noun for something.

a rant of talk radio hosts —@maineone

a shutter of tourists —@itdoug

a drama of teenaged girls —@kchichester

a pew of fundamentalists —@reganref

a pack of smokers —@rissriss

a clot of hematologists —@jtopf

a slouch of teenagers —@samjwatkins

a sack of recently unemployed persons —@Sandydca

a smarm of newscasters —@arielleeve

a quibble of analysts —@jayfry3

a slick of Apple products —@cloud64

a drove of cars —@equinox_child

a scowl of Republicans—and to be fair, a futz of Democrats —@passepartout

a squeal of Jonas Brothers fans —@kimwolf

Invent a witty collective noun for something.

a mass of marketeers —@BostonBrander

a torrent of software pirates —@mazola_jr

a guzzle of SUVs —@CMich

a crock of right-wing pundits —@BobFenchel

a tussle of toddlers —@ortrek

a rash of dermatologists —@mimbrava

a battalion of armadillos —@relsqui

a disdain of vegans —@hatchethead

a clutch of purses —@hriefs

a shush of librarians —@erichines

a snap of college a cappella singers —@jager13579

a mass of nuns —@hriefs

a mess of 2-year-olds —@hriefs

a hype of Trumps —@cswriter

a slither of politicians —@noveldoctor

a sulk of teenage boys —@bphuettner

a Kutcher of followers —@noveldoctor

Heard any good puns lately?

Zen master goes up to the hot dog vendor and says, "Make me one with everything." —@McConnellpdx:

What do you call a group of rabbits walking backwards? A receding hare line. —@Telp:

I couldn't remember how to throw a boomerang. Then it came back to me. —@marianneoconnor

I once spent thousands of dollars on a reincarnation seminar. Figured what the hell—you only live once! —@kw_sendtonews

Q: What do you call a deer with no eye? A: No eye-deer. —@whataboutbob

A man walks into a bar with a slab of asphalt under his arm and says: "A beer please, and one for the road." —@tmasteve

I dated a trapeze artist once, but he dropped me for someone else! —@KillerMango

Do my taxes on my computer? Nah. I'm just not that Intuit. —@wide_awake

The Second Amendment gives Michelle Obama the right to bare arms. —@animaliac

She was only a moonshiner's daughter, but I loved her still. —@MatthewDooley

Did you hear about the new line of Elvis Presley–themed steakhouses? They're for people who love meat tender. —@mmcd908

Didya hear about the woman who backed into a propeller? Disaster.
—@Dougwalk

Eskimo out fishing; gets cold, builds fire, sinks, and dies. Moral: You can't have your kayak and heat it too.
—@mmcd908

I heard a guy play Dancing Queen on a didgeridoo. I thought, "That's aboriginal."
—@marchdigital

Q: What do you call someone else's cheese? A: Nacho cheese. —@gavman678

Show me where Karl Marx is buried, and I'll show you a communist plot. —@same

Pogue Sez

Before this whole book thing started, I had been posting on Twitter, just for fun, a groaner pun every night. I called it "Tonight's Meditation."

Soon enough, it became a tradition for my followers to try to top my pun—and one night, this tradition led to a veritable punathon.

It began when I passed on a pun sent to me by @killermango: "I dated a trapeze artist once, but she dropped me for someone else."

At that point, my followers began continuing the trapeze-love affair story with a series of ever more ridiculous puns: "We met on a singles bar." (@BrianWolven) "She really threw me for a loop." (@yuetsu) "I've never fallen so hard for a woman in my life." (@whistlingfish)

"I was head over heels for her, but alas, I caught something." (@JimFL) "She wanted to be in the spotlight all the time." (@xtello) "We kept going back and forth. Later, I found out that she was into swinging. Net, net, I'm better off without her." (@JimFL)

And finally: "Maybe I'll get lucky on the rebound." (@JimFL)

Wonder if the *Guinness Book of Records* has a category for Longest String of Lame Puns?

Two hydrogen atoms meet. One says, "I've lost my electron." The other says, "Are you sure?" The first replies: "Yes, I'm positive." —@siouxt

I thought of almost a dozen puns that would tickle your fancy, but alas, no pun in ten did. —@vidiot_

Can February March? No, but April May! —@armenoush

Cattle rancher dies, leaves the ranch to his sons, who decide to change its name to Focus—where the sons raise meat. —@hoellars

Two TV installers met on a roof and fell in love. The wedding ceremony wasn't so great...but, wow, the reception!! —@marque0

I tried to teach my horse philosophy, but it just didn't stick. Never again will I put Descartes before da horse. —@zwb

If it's not one thing, it's your mother. —@harold_cypress

I stood there wondering why the Frisbee kept getting bigger. Then it hit me. —@mindofchester

GM has announced a car called the Amnesia. It's the only car they make that they can't recall. —@kcheriton

Q. How long does it take to do a pirate's income tax return? A. About an arrrrrr. —@librarychip

Opening a new funeral parlor can be quite an undertaking. —@nniuq25

I've been reading Stephen Hawking's latest book about anti-gravity. I can't put it down. —@ianw91

Two cannibals are eating a clown. One says to the other: "Does this taste funny to you?" —@peterdjohnston

One cannibal remarks: "I don't like my mother-in-law." The other one says: "So try the soup!" —@DCDawg

I heard a guy got into an accident and had his whole left side amputated. He's going to be all right. —@steveamiller

What does a dyslexic agnostic with insomnia do? Lies awake at night, wondering whether or not there really is a dog. —@jakerome

When I asked if I could get insurance if the nearby volcano erupted, they assured me I would be covered. —@legweak

Two parrots sat on a perch. After a moment, one said to the other: "Do you smell fish?" —@alpalmer

This just in: A hole has been found in the nudist camp wall. The police are looking into it. —@nickfruhling

Don't buy flowers from monks. Only you can prevent florist friars. —@dschach

I just got rid of my waterbed. My girlfriend and I were drifting apart. —@sfmarc

A vulture boards a plane carrying two dead raccoons. Flight attendant: "Sorry, sir, only one carrion per passenger."
—@UNnouncer

Those who forget the post are condemned to retweet it.
—@supertweeting

Those who forget the pasta are doomed to reheat it.
—@mrzwiggy

The bride had just three thoughts as she entered the church to get married: "Aisle...altar...hymn!" —@evan1t

Did you hear about the two tanker ships that crashed? One carried red paint, the other purple. Now both crews are marooned. —@SharonZardetto

A man fell into the reupholstery machine. But it's okay—he's fully recovered. —@StubbornlyWrite

WARNING: Incorrigible punster on the loose. Do not incorrige him. —@oxfordcleric

Q: What did General Google say when confronted by an army of loyal twitterers? A: Retweet!! —@Trick_or_tweet

A great pun is its own reword. —@davemark

Caption this photo.

Really? That's interesting...MY family has 2.5 kids, too!
—@pumpkinshirt

So can we claim two dependents on our taxes now?
—@ColinDabritz

Hey...No one told me I had a half brother!! —@Biha

I knew the "Half Off Superhero Costumes" sale was too good to be true. —@danagel

Uh-oh. Mom's going to know for sure that I was playing with her Ginsu knives. —@etreglia

Finally! Two cases that will stump Dr. House—an upper case and a lower case. —@brianwolven

The Amazing Spider-Man! Half man, half...other man. —@2dkid

Paulie Parker was only half the man his older brother Peter was. —@MacRtst

The divorcing parents worked out a unique joint-custody arrangement. —@clneeley

Because I'm the half with the brain, and I say so! —@pumpkinshirt

AHA! You missed me!...Oh dear. —@HayleyLovesYou

The Green Goblin strikes again! —@_B_en

Who's got the worst romantic-dumping story?

Valentine's. Candles, chocolate, champagne. Soft kisses, searching my eyes: "God told me who I'm going to marry, and it's not you." —@WPTunes

She came to visit for a long weekend with my extended family. Broke up Thurs night, then stayed till Tues morning. —@designerbrent

My ex-husband tattooed his girlfriend's face onto his bicep two months after meeting her, BEFORE telling me anything! —@griffindl

By telephone, 10 days before our wedding. Everyone who knows me is sick of this story by now. —@behindthecamera

My boyfriend was dumped by his ex in the delivery room while giving birth to their son. —@beth910

After four years together, my ex dumped me via Yahoo Instant Messenger.
—@jenconnic

New Year's Eve at midnight, right after the kiss. —@TheUndomestic

Call from girlfriend: "My brother told me I should break up with you. Don't argue—he's in the Mafia." —@noveldoctor

Pogue Sez

Part of the fun of putting together this project was "retweeting" the five best responses to each question.

At about 10:30 p.m., I'd ask my followers a question; by 10:31, the responses would begin pouring in. Over the next thirty minutes, I would pass on (retweet) the best responses to my entire group of followers.

When retweeting, you use a particular format. For example, if your Twitter name is @Bongoman and you send me a pun—"A good pun is its own reword"—I'll pass that on like this: "RT @Bongoman: A good pun is its own reword."

In other words, I'm saying: "This is Bongoman's thought; I'm passing it along."

Newcomers to Twitter, however, don't know what RT means—and they often became hilariously baffled by the rapid succession of retweets for this book.

When I passed along five of these horrific romantic-dumping stories, for example, several Twitter beginners assumed that all five came from my *own* romantic history. They were aghast.

"Dude—you've had the worst luck w/women of anyone I've ever heard of!" wrote one. "Better luck next time!"

"So sorry," said another. "I promise not all women are like that!"

My sister found out her boyfriend was cheating when she hit play on his VCR and saw the video evidence.
—@Dreadkiaili

At a wedding where I was a bridesmaid, I overheard him using the pay phone to make a date for the next day. (I didn't catch the bouquet.) —@elizabeth627

My ex went out to get the ol' pack of cigs—and never came back. Bastard.
—@EJMaranda

My girlfriend in high school came to where I was working to collect her hair scrunchie I kept on the shifter in my car. I knew it was over. —@bnl771

Driving LA→SF. Car dies. $500 + 4 hours to fix car. Drive 2 more hours. Car dies again. $100 tow to hotel. She dumps me at hotel.
—@dsrichard

1993: While being dumped by girlfriend, phone rings, machine picks up, and her new boyfriend leaves message that I get to hear. —@dsrichard

Me, posed in lingerie, sexy music, candlelight, come-hither look... husband walks in, rolls eyes, says, "Not again!" and walks out. —@shelleyryan

My girlfriend was showing me an appointment book she got for Xmas. Scheduled in for that day was, "Break up with Jay." Guess my name. —@jadawa

He said I was too young for him. (He was 27; I was 20.) The next day, he IM'ed my 19-year-old roommate to ask her out. —@demetria23

Facebook Newsfeed: My girlfriend's status was now listed as single. —@JHKramerica

Guy told me he was going to marry another woman in 1 week...and he had to do the right thing. (Yeah, mighta been awkward.) —@VenetianBlond

Dumped in high school, rode subway (NYC) home in tears, train ran over homeless person, stranded in the Bronx. —anonymous

Schlepped half my worldly possessions from ATL to girlfriend's in D.C. Was getting the other half when she phoned to say she cheated. —@stevenaverett

Drove 8 hours for Valentine's weekend. Made out a lot. Her before I go home: "I have a 5-year plan, and you're not in it." —@bahamat

Guy I was seeing blogged how he wasn't looking for a relationship. And all my friends saw it. —@helenahandbskt

Came back from summer break for sr year of college. Someone congrat'd me; he'd heard my girlfriend had gotten married. True, but not to me! —@danalog

A girl broke up with me because she couldn't see us together in the future. She and her "life partner" now make organic picnic baskets. —@Notsewfast

I was probably the first person ever dumped via e-mail. In 1994. —@mweintr

He chose 1st qtr of playoff game that went 6 qtrs to say "no more." I cried 2 hrs while we made love 1 more time. He watched the last qtr. —@tanabutler

On my honeymoon. At least there was champagne... —@redstatestephen

She moved with no forwarding address. —@InfiniTom

Let's hear your best marriage-proposal story.

My best friend Todd got Billy Joel to speak the proposal during a concert with Elton John at Rich Stadium in Buffalo, NY. —@GrumpyHerb

She worked at Newsweek, me at Time. I made a fake-magazine/proposal, and Book Court placed it on magazine rack minutes before we walked in. —@eroston

I proposed on a banner on a St. Paddy's parade float. She was mad, ready to dump me. I pointed. She: "So?" Thought it was for someone else! —@McCArch

Met my wife, & 10 minutes later told her she'd marry me. Her reply: "You're crazy!" Me: "Yes, but you'll still marry me." Wed 1 year later. —@rkzerok

Going out 6 weeks. Her: "This risotto is so good you could pop the question." Me: "Brilliant idea!" So I did, on one knee. Her: Gobsmacked! —@vonschlapper

I proposed by videotape, but then left town to attend a silent retreat (no phone) for four days! —@scottydc1

My wife was so busy getting ready to go out that I couldn't get her to hold still for me to propose. I had to corner her in a closet. —@hanineal

My wife (after 6 glasses of wine in a 12-course meal), after finding a diamond ring in her dessert: "Hey! Someone put a ring in here." —@hhlodesign

My cousin (F) proposed during a hot-air balloon ride. Told him they weren't landing until he said yes. —@DixieJo719

Aunt teaches little kids. Uncle dressed up as Barney for in-class party. Popped the question in front of class in costume (w/out headpiece). —@mediasmaven

Dolphins at aquarium, private session. Trainer: "Do you know what the dolphins are saying?" He said they were telling him to ask me to marry him. —@osxgirl

My dad proposed to my mother at Denny's. They've been together 27 years. —@thewesterly

My son took his girlfriend geocaching and secretly hid the ring among the plastic trinkets in the treasure box. —@Sandydca

I saw a proposal up on a menu chalkboard at a restaurant. The happy couple bought a round of drinks for everyone. —@pattyfab

The night he proposed, my husband sent a limo to pick me up from work. But I thought it was a scam at first and refused to get in! —@kellyturner

I buried the ring box in the fossil-discovery sandbox at a dinosaur museum. We dug for it. The kids freaked out! —@RyanMoffitt

Billboard for diamonds:, "Amy, marry me!" It was just an ad, but my friend Amy called Glen and said, "Yes, I will!," so he went with it. —@holle8e3

Dusk. Huntington Beach. 250 tea-light candles, spelling out, "Elizabeth Marry Me?" It was epic. —@postphotos

Signal flags on USS Constitution in Boston..."Marry Me -Karin"...She said yes! —@jotulloch

I proposed in my newspaper column. —@DaveLieber

I had our SCUBA instructor plant a treasure chest with the ring & some costume jewelry at the bottom of a lake we were training in. —@caughill

My friend planned an elaborate proposal at a state park in another city—then chickened out. Next day, he popped the question at a Wendy's. —@pumpkinshirt

My cousin's fiancé wanted to propose in Hawaii; airport security asked him to open the dubious box. Had to propose on the spot! —@vikaman

My boyfriend flew me to Paris for Valentine's as a "Sorry, I'm not ready to propose" trip—proposed in a little park near the Eiffel Tower. —@thinkc

Dental hygienist: "How'd he propose?" Me: "Mm, he didn't really!" Hygienist dragged us, bibs on, to waiting room, to make him ask properly! —@mjfrombuffalo

I proposed via a painting placed in a gallery on gallery-walk night in Milwaukee. The story won us a dinner on our honeymoon. —@streetzapizza

Guy gave a girl a puppy with a ring attached to the collar. Girl didn't know which was more exciting—the puppy or the new fiancé! —@nighttime_star

I proposed at Halloween by carving "Will you marry me?" in a pumpkin. —@shaiberger

When my brother asked my now–sister-in-law, she immediately said yes...and then threw up because she had been so nervous. —@thebandnork

I know a couple who got engaged during an argument. "WELL, WHY DON'T WE JUST GET MARRIED THEN!?" "FINE!" "FINE!!" —@relsqui

You know the honeymoon's over when...

...you find your copy of "The Joy of Sex" marked 50 cents at your garage sale. —@rponto

...he calls home to say he'll be late for dinner, and the answering machine says it's in the microwave. —@Eis4everything

...if you get tied to the bed, you're left there for days. —@Fejennings

...you quit blaming the dog. —@lizardrebel

...Honey? Can you please trim the corns on my feet? —@_B_en

...your spouse turns to you and says: "You don't spend enough time out with your friends." —@egreshko

...you get Home Depot gift cards for your birthday. —@rponto

...you find Weight Watchers coupons left on your pillow. —@rponto

...the bachelor stuff she let you keep is now in a "donations" box by the door. —@Biha

...you reach over to the next pillow for a hug and your dog licks your hand. —@williammarydoc

...you look forward to going to the movies together and seeing different films. —@hriefs

...Three words: Bathroom door open. —@Stefaniya

...the dog has more "pet names" than you do. —@JimFl

...every time your wife passes through the room, she turns down the music. —@williammarydoc

...instead of sexy lingerie, he gives you a sweatsuit for Christmas. —@Dolores_Rosario

...both of you stop shaving. —@passepartout

...the only action in the back seat of your car is installing a car seat! —@_B_en

....the toilet seat is taped shut. —@davidmorar

...date night = Taco Bell. —@_B_en

Write an amusing personal ad for somebody famous.

MWM, father of country, seeks companion for cherry tree-cutting excursion across Delaware River. See $1 bill for photo. No Tories. —@CajunBrooke

SWM ISO congregation to lead. Must be able to live 40 yrs in desert. Love of manna a plus. Only 10 easy rules to follow. —@jschechner

Illuminated, soul-seeking, high-voltage relationship. I'm a real "dot the 'i's" kind of guy. Contact Luxo Jr. @Pixar —@rponto

Man made from Tin seeks working heart. Will pay extra for oil change. Meet along yellow brick road. —@EvanFogel

FOUND: One glass shoe for tiny foot. Owner last seen riding off in a giant pumpkin at stroke of midnight. If fits, will share kingdom! —@bfulford

Juneau babe seeking educated man for some tutorin' and edu-catin' on all this international stuff. Contact Sarah2012. —@mkh531

SWF seeks SWM from other side of tracks for friendship, love & more. No drug users, please. Call Verona416, ask for Juliet. —@vonschlapper

SBM 7'1. Looking for lucky #20,001. —@richardabailey

Recently retired head of state is back in TX. Looking for friend-ship and good times. Anyone? Anyone? Please? —@DrewJazz

Looking for that certain someone—call Heisenberg at 2 or 3 or 2. —@kvetchguru

PM seeks woman for late nights, champagne, baths, cigars, Tory politics. Experience with bomb shelter a plus. Write wsc@10downing. —@CapitolClio

Young boy with magic in his eyes, seeks equally powerful girl with a thing for scars. —@XmasRights

SWM seeking SF. Enjoys paint-by-numbers and vivid landscapes. A bit hard of hearing but a catch in his warm yellows phase. —@thumbble

Wanted: SWM seeks horse for my kingdom or best offer. Contact Richard III. —@msewall

Sea captain seeks Caucasian BBW (big beautiful whale). Need someone who won't drag me down. —@BaronSchaaf

MBM seeks short-term (up to 8 years) relationship with superpower for restoration and fun. Must love dogs. —@grousehouse

Man ISO opposite for clothing-optional community. Gardening a plus; no apple lovers, pls. Call 1; ask for Adam. —@ProfJonathan

Lonely N Korean male, enjoys wearing chicks' glasses. Relieve stress by popping off nukes. It's how I roll. Kim, illy, it's all good. —@loudstone

Omnipotent entity seeking same. Must be monotheistic. —@jadawa

Make up a phony Chinese proverb.

Salt is needed to taste the sugar. —@vdr

Soup is best eaten with a spoon. —@Boxianna

When water is flowing uphill, something terrible is sure to follow. —@jschechner

Romance dances all night; love massages the corns at dawn. —@pumpkinshirt

If a bird can't fly, it walks. —@benjamin_gray

Give a man a fish, he will eat for a day. Give him two fish, and he will put together a complex scheme of fish derivatives. —@pumpkinshirt

It is dangerous to leap a chasm in two bounds.

—@DaVeroFarm

Even a gentle breeze will litter a king's lawn. —@gadruggist

The smartest dog still begs from children. —@pumpkinshirt

If a man wants to grow a long tooth, he should have the lip to cover it. —@rjampro

Squeezing a dry sponge never yields water. —@ericgrajo

A shield of angry words has a knife for a handle. —@pumpkinshirt

Milk the cow, but don't pull off the udder. —@hvddrift

Odes to the bear are seldom written in the woods. —@pumpkinshirt

Deprive any mirror of its silver, and even the czar cannot see his nose. —@eboychik

The cynic is a thirsty ghost in a wine cellar. —@pumpkinshirt

A swarm of houseflies will never make honey. —@whistlingfish

When the songbird dies, his song also dies. —@davidvnewman

A mouse may live in a castle, but it cannot open the door. —@pumpkinshirt

Man who eats chicken for dinner won't have eggs for breakfast. —@SubtleOddity

When the peacock shakes his tail, everyone pays attention. —@davidvnewman

The pit is always smaller than the plum. —@yodaveg

The man who looks up and trips still sees stars. —@pumpkinshirt

The young man shades the seedling; the tree shades the old man. —@cherimullins

One bitter truth is a banquet; a thousand sweet lies but a morsel. —@MrMadman

Strike the devil in fury and, beaten, he wins. —@pumpkinshirt

One person talking at once is speech. Ten people talking at once is noise. A million people talking is silence. —@TonyNoland

The man who quotes himself does credit to neither. —@pumpkinshirt

If you close your doors in difficult times, then no good can come in. —@sgbulfone

Arrows may be called back, but seldom do they answer the call. —@pumpkinshirt

Bend the knee and the waist before the fingers. —@pumpkinshirt

A lamp on a street is worth ten guards. —@pumpkinshirt

A pig with a cold still makes good bacon. —@beuktv

What's the best present you ever received (besides your kids)?

A very ugly water-filled vase with plastic red flowers from my 8-year-old son for my birthday (with his own money). —@plocke

A timeline of my life, hand-drawn by my 12-year-old sister, for my 25th birthday. Creative and personal. Almost 20 years ago. —@GrumpyHerb

I was expecting a train set for Xmas, didn't get it, hid my emotions. Dad asked me to fetch wine downstairs—just so I'd find the train set. —@convagency

When I was wee small, my mom, the non-crafts person, secretly knitted me a Very Long Blue Stocking Cap for Xmas. Hideous. Perfect. —@marthasullivan

A Paul Westerberg electric guitar for Christmas two years ago. My parents would never get me one when I was little. Then, at 34, bingo! —@kaline

Went out Christmas morning to find a shiny new RX-8 from my wife sitting in the driveway. Even had a giant bow on it! —@Roger_Nall

A bag of used paperbacks from a street seller, all chosen with care by my husband. Many hours of escape when I needed them. —@lirot

My parents gave me $5,000 cash to splurge on stuff when I bought my first home. —@bnl771

My grandmother's engagement ring, sized to fit me. —@bfulford

A new Hummer H2. (Please publish without my name.)

1st gift my kid bought with own money: mug, choc-covered espresso beans, gift card from fave café. A secret: I'd stopped drinking coffee. —@bekSF

A picnic quilt, sewn by my aunt, of squares cut from old Levis pant legs. Useful, indestructible, and beautiful. —@foolery

Lost my wedding rings in January. Hubby bought me a new expensive set for our anniv in April. Found my rings 3 weeks later. (He's not mad :) —@Laurita369

My dad gave me a 486 motherboard when I was 19. Launched my career. —@bahamat

Two: an Easy-Bake Oven and a promise ring. Different times, equally as excited. —@indiereads

Husband composed a song for me and surprised me at a concert by playing it with the band. —@Dreadkiaili

'71 Caprice Classic 2-door from my parents when I got my license. My responsibility to upkeep. Responsibility: BEST. GIFT. EVER. —@sassyscorpio71

A stranger mailed me a picture of my father at Forbes Field, shaking Roberto Clemente's hand. —@MaryHr

19th birthday: Mom wrote to dozens of celebrities, gave me a binder filled with personalized autographs from celebs saying Hi! —@gfitzp

1993 GT DynoCompe BMX bicycle. My single-parent mom picked up a third job so I could have it for my birthday. Never forgot it. —@Purnell

He took me to Isle of Palms, near Charleston, SC, where he had rented a house for the weekend and filled it with my friends. A surprise. —@citygirlgvl

A friend of mine let me drive his Ferrari Enzo, therefore ruining all future birthday presents for the rest of my life! —@nocturne1

One brother flew cross-country, and the other drove seven hours, just to take me to dinner on my birthday. —@LADinLA

A comic strip about them forgetting my birthday; they let me fill in the last frame. —@hobbesdream

A signed copy of "So Long, and Thanks for All the Fish." —@jadawa

I was in a drive-through, and the stranger ahead of me paid for my order for no apparent reason. —@CreepyGnome

My wife surprised me and arranged for a gospel choir to sing "Happy Birthday" to me. Wow, was I shocked! —@tjredbird

My boyfriend gave me a giant wooden clothespin. It is the oddest, coolest thing ever! —@mkovo

10 years ago, my mom gave me enough money so I could come up with the down payment and buy my house. —@GardenGirl1214

The green jacket my golfer father gave me after I received my master's degree! —@macnbc

Extra cable TV package for 6 months, so I could watch all the Pittsburgh Pens hockey I wanted while living in MN. Unexpected. Perfect. —@thigpensrevenge

A Gmail account, back when you needed an invitation. —@dbrewer80221

A toaster oven for Mother's Day, from the son who always forgets. It could have been anything, and it still would have been the best. —@bornfamous

My girlfriend surprised me with round-trip tix to St. Louis to see a Cardinals' game from the Commissioner's box. Perfect! —@sodajunkie

I woke up to a hot-air ballon outside my bedroom window (I was 16). Got in & floated above Phoenix! —@afairfield

A Rolex, for my birthday, that was made in my birth year. I never take it off. —@hriefs

The best present I ever got was my first bicycle when I was 8 yrs old—it meant freedom! —@RayPrevi

Make up a clever title for a sequel to a famous movie.

12 Angrier Men —@ysteinbuch

When Harry Left Sally —@ltlachiever

Chitty Chitty Bang Bang. Bang. —@SteveKlinck

Wall-F —@rponto

Star Wars XXVII: The Empire Goes Chapter 11 —@Enclarity

Dinner at Cartier's —@lizardrebel

OK, The Postman Sometimes Rings 3 Times —@bnl771

The XI-Men —@edwardtwoo

Rachel Getting Divorced —@lizardrebel

Star Trek: The Search for Star Trek —@REReader

The Morning After the Night at the Museum —@AndyMaslin

Die Another Day Another Day —@HawkW

Charlie After the Chocolate Factory: The Diabetic Coma —@lizardrebel

Even the Devil Can No Longer Afford Prada —@Sandydca

Make up a clever title for a sequel to a famous movie.

Debbie Does the Greater Dallas/Ft. Worth Metroplex —@ajds

Tropic Thunder: Arctic Breeze —@davidsillen

Pretty Old Woman —@vonschlapper

My Big Fat Greek Divorce —@angeldominguez

Monsters, Inc.: Chapter 11
—@artemis822

Sophie's Second Choice —@ronhitchens

Bob & Carol & Ted & Tom —@hriefs

Das ReBoot —@DivorcingDaze

Lord of the Ringtones —@vonschlapper

My Dessert with André —@brianpjcronin

It's a Wonderful Afterlife —@tepeka

Sex, Lies, and YouTube —@rocketsurgery

Monsters vs. Aliens vs. Predator —@FredTweetzsche

Alienses —@acinonnap

The Neverending Story: The Final Chapter —@tommertron

The Post-Graduate —@Marcodj

Never Say Never Again Again —@evan1t

Malcolm XI —@kurometarikku

Twistest —@BarrSteve

Flushing Nemo —@erikmitk

Old Frankenstein —@RichardMarini

Lowered Expectations —@ajds

Born Again on the 4th of July —@mmarion

The Day After the Day After Tomorrow —@markowitz

There's Something Else About Mary —@reganref

What's the best bumper sticker you've ever seen?

Witches' parking. All others will be toad —@AlanEdinger

Jesus loves you, but I'm his favorite —@mviamari

Stop Plate Tectonics Now! —@mjf2009

Your child may be an honor student, but YOU are still an idiot —@jmartellaro

Sorry to be driving so close to the front of your car —@anagramfx

Ask me about my crippling social anxiety —@havochaos

I married Ms. Right. Her first name is Always. —@alfamale1RT

My kid can beat up your honor roll kid —@rkaika

Curb your God —@toshkin

Dyslexics Untie! —@mhhughes

Silence is golden, but duct tape is silver —@christhespy

If you can read this...I've lost my boat! —@weta

Dog is my co-pilot —@Marcviln

I bet Jesus would have used HIS turn signals —@rocknorris

Filthy, stinkin' rich. (2 out of 3 ain't bad.) —@kahugary

I Park Like an Idiot. (I keep a few of these with me, just in case.) —@edgeofwithin

I am nobody. Nobody is perfect. Therefore, I am perfect.

—@HarrisonWilson

Don't be pessimistic. It probably won't work anyway. —@martinvipond

Jesus Loves You. He's just not IN LOVE with you. —@vikaman

Pontius is my Pilate —@DVDGeeks

VEGETARIAN: Indian word for poor hunter —@PatrickDuany

Jesus Saves!...Esposito goes in for the rebound. —@chrisreich

Buckle up. It makes it harder for the aliens to suck you out of your car. —@sheppy

You—Out of the Gene Pool! —@MILewin

LIBRARIANS: The original search engine —@maggda

Jesus WAS my copilot...but we crashed in the Andes and I had to eat him —@DrKoob

WARNING: I drive like you do —@chrismtp

Welcome to Berkeley. Here's your bumper sticker. —@rotomonkey

As long as there are tests, there will be prayer in public schools —@mehughes124

When everything's coming your way, you're in the wrong lane —@pfrankly

Due to the current financial restraints, the light at the end of the tunnel will be turned off until further notice —@andrewall

What Would Scooby Do? —@sylvainbbb

Local 838, unemployed English majors' union —@pamela_gill2000

Keep honking while I reload —@MBOZE

(On a trailer carrying sheep): Ewe Haul —@aweber9

Get any closer, and I will flick a booger on your windshield —@llennon

Yes, this is my truck. No, I won't help you move. —@sgallow98

Honk If I'm Paying Your Mortgage —@tcervo

JESUS LOVES YOU. The rest of us think you're an asshole. —@pik0

Driver Carries No Cash—He's Married! —@jciesielski

Jesus Is My Car Insurance —@RHere2

Republicans for Voldemort —@skishua

Illiterate? Write for this free booklet! —@Woogel

(Next to upside-down Nike logo): Just did it! —@gafguru

JESUS IS COMING. Look busy. —@grabbingtoast

My karma ran over your dogma. —@sgoodin

What spoken phrase drives you crazy?

"Promise you won't get mad." How can I promise if I haven't heard it yet?! —@HyunINC

"Well, anyhoo..." (It was cute the first THOUSAND times...) —@vdr

"Cold as hell!" —@dougredding

"Moving forward." Who says "Moving backwards?" Though people who talk like this may be moving "sideways." —@DrQz

"by and large." Why does this mean, anyway? Where did it come from? —@ruhlman

Redundancy is highly annoying: ATM machine. SAT test. PIN number. Tuna fish (as opposed to what—tuna cow?). —@jschechner

"It's all good." —@ejinla

"It's a quantum leap in..." Why such a small leap? —@heartydiamond

Said during times of stress or annoyance: "Ya gotta laugh." No, no, I don't. —@saintjimmy

"Irregardless." Huh? —@WaterGeek

"Where are you AT?" What does the "at" add, other than a show of poor education? —@jschechner

"No problem." Why would handing me my purchase and change be a problem? —@AllenJM

"Shut UP!" Said when the listener actually means: "Tell me MORE..." —@maryegilmore

"Needless to say..." Then DON'T say it! —@c173

My all-time favorite, pure corporate drivel: "We're taking it to the next level." Which level is that? Is it up or down? —@hartfortworth

I hate it when you ask someone, "How are you?," and you get, "Real good." As opposed to "Fake good"? —@godlikescoke

"How are you?" —"Can't complain." AAARGH! You just DID, with that passive-aggressive answer! —@BoffleSpoffle

"Conscious decision." When are decisions not conscious? —@bxchen

"A fraction of the cost." Isn't 99/100ths a fraction of the cost, too? —@podfeet

"This [insert situation] has gone South." Why can't things go North when they don't go as planned? —@66sweep

This one makes me grind my teeth: "It goes without saying." —@aceflyrite

"Cool beans" is very annoying. —@pinlux

"May I interrupt you?" Uhhh....you just did... —@egreshko

How about when people agree with you and say, "I don't disagree"? —@Schwartzie14

"Multimedia." (The word "media" is already plural!) —@jeffchr

"__ is very unique." Doh! "Unique" already means one of a kind. It cannot be modified. —@jkennett

"With all due respect..." Right before people show absolutely no respect. —@Jaste

In a face-to-face group meeting: "Let's take these questions offline." How would we get MORE offline—die? —@drewallstar

"Disconnect." Since when did this verb become a noun? What's wrong with "disconnection," which served OK for years? —@cloud64

"Pre-plan." So you're planning to plan!? —@D_Chan

"I'm dubious" drives me crazy. I want to yell, "You ARE dubious! But you probably meant, 'I'm doubtful'!" —@yacitus

"Too clever by half." I'm not sure I even understand the mathematical formula on that one. —@UncJonny

News reports of "a brutal murder." Are some murders benevolent? —@JazzPreacher

How about people who say they are going to "unthaw" something? Really? So you are freezing it, then? —@MeaganRhoades

"Going forward" drives me nuts, and it's used at nearly every meeting. Unless you've got a time machine, where else are we going? —@iDoogie

"The thing is, is that…" —@jenny_blake

"Yeah, no." What the hell does that even mean?? —@aulia

"I could care less." It's "I COULDN'T care less." If you COULD care less, go ahead and care less! Grrrrrr! —@Sandydca

"No worries." Are we all on a tropical island? —@pjdempsey

"I'm not a happy camper." (Why can't people be unhappy without camping?) —@floretbroccoli

"What can I DO you for?" (Unless you're in overalls working at the General Store, don't ask me this.) —@gdouban

Caption this photo.

New ride at Six Flags over Iraq. —@tmasteve

But Mommy, I don't WANT to go visit Grandma and Grandpa! —@donnaaparis

After the circus stuntman retired, they had a hard time finding a replacement of the same caliber. —@shayman

Sadly, another fodderless child. —@markowitz

When I was a boy, I WALKED to school... —@jkennett

Billy's father pointed him in the right direction. —@stenro

From then on, hide-and-seek was forbidden in the Smith household. —@maggie162

Ryanair: Lowest-cost fares. —@JuanluR

Conclusive proof that the North Korean nuclear program is not as advanced as originally thought. —@larsperk

Johnny misunderstood what his dad meant by, "Who wants to be first to be shot by my new Canon?" —@dudgeoh

Make up a clever title for a prequel to a famous movie.

The Day Before Tomorrow —@peterluna

Flirts with Wolves —@DSpinellis

Mr. Smith MapQuests Washington —@michaelbuckman

Singin' in Potentially Inclement Weather —@dvdmon

Irritated Bull —@pabaker55

Matchbook of the Vanities —@KlarkKent007

Grapes of Slight Irritation —@tsmyther

Trouble Crossing the River Kwai —@pabaker55

Young Yeller —@tsmyther

The Way We Are —@mostlymostly

Rachel Getting Engaged —@donnadalewis

Field of Corn —@ChrisNBC13HD

Indiana Jones and the Missing Wallet —@TonightWeParty

The Undergraduate —@metallicajake

Make up a clever title for a prequel to a famous movie.

Hitchhiker's Guide to the Lower 48 —@kbranch

The Stepford Fiancées —@Jefficator

Dining with the Enemy —@Jefficator

Finally We Are in Las Vegas! —@nsnigam

Medical Student Zhivago —@lizardrebel

Chess Club: THAT Should Relieve Our Aggression! —@EvanFogel

Close Encounters of the First Two Kinds —@TaxMan45

The Poets' Society —@larrymustard

The X-boys —@MisterRoo

Snakes in the Terminal —@justinchambers

A Grape in the Sun —@mtsullivan

Annoyed Calf —@michaelbuckman

Apocalypse At-Some-Point-to-Be-Determined —@atomlinson

Conceived on the 4th of October —@jadawa

Debilitating Muscle-Twinge Mountain —@gskull

Drinking Coffee Late at Night in Seattle —@JaredParker

Foreigner Kane —@chrisw427

Heavy Petting in the City —@mldrabenstott

Monsters Startup —@satya893

Chariots of Highly Combustible Materials —@Urban_Orchards

Night of the Living Living —@euskir

Saturday Night Sniffles —@michaelbuckman

Star Sanctions —@afairfield

We're Running Low on Mohicans —@rllewis

The Devil Wears OshKosh B'Gosh —@gskull

The Maltese Hatchling —@Ophelia673

The Manchurian Petition —@michaelbuckman

Two Dalmations and a Fertility Clinic —@MikeGluck

Violinist on the Porch —@ljhardin

Just A Little Weird (prequel to Psycho) —@smccormack

Blind Date with Wolves —@yardsalesgalore

Planting Arizona —@amorak

There Goes Private Ryan...I Hope He'll Be OK —@slightly99

You know it's time to look for a new job when...

...you're asked to take a business trip to Mexico at the height of the swine flu outbreak. —@hriefs

...you see YOUR job listed on Monster. —@AndyMaslin

...even telemarketers for toner suppliers stop calling your company. —@MediaOwl

...the bosses start saying things like, "Do more with less." —@Rosalitagirl

...suits stop by your cubicle and ask, "Are you still here?"

—@ann_walker

...you get the company's new org chart...and you're not on it! (Actually happened to me.) —@BlindscomCEO

...you begin Googling for instructions on getting your name INTO the jury pool. —@ImageSpecialist

...the lobby security guards point and snicker as you enter the building. —@hriefs

...every day, you feel that you're living a Dilbert cartoon. —@hriefs

...Obama has a press conference in the rose garden about rescuing your company! —@rylanhamilton

...you get your "25 years of service" watch, and you're still working in the mailroom. —@jazcan

..Mike Rowe is the only guy representing your struggles. —@ElePhatt

...a trip to the dentist for a root canal is welcome time away from the office. —@jodisc

...people start sticking their heads in your cubicle, pointing at something, and yelling, "Dibs!" —@tom_streeter

What's the most memorable pickup line you ever heard?

Is it hot in here, or just you? —@AlanHowlett

The only thing your eyes haven't told me is your name. —@babygeniusgirl

Frequently heard during my waitress days: I'll have a waitress on toast—with no dressing. —@bornfamous

I'm thinking about becoming a pirate so I can get some of your booty. —@flaminglawyer

I know that milk does a body good, but dang, girl...how much have you been drinking? —@jormadotcom

Do you have no left arm and no left leg? Because, girl, you ALL RIGHT! —@richjlee

You must be a parking ticket, because you've got fine written all over you.

—@EvanFogel

How do you like me so far? —@lrinner

Do you have the time? [Gives the time] No, the time to write down my number? —@sarahbeery

You are attractive and busy, and I'm attractive and busy. So... let's not waste each other's time! —@SexySEO

You won't be able to break my heart, because it melted when I met you. —@Sky1Ron

I'm writing a phonebook; can I have your number? —@laurentmeyvaert

You got a match? (No.) Then how 'bout you and me? —@PowerLunch

I wish I was your derivative, so I could lie tangent to your curves. —@Theatergirl62

Do you sleep on your stomach? (No.) Can I? —@Larrco

We should do breakfast tomorrow. Do you want me to call you or nudge you? —@DavePurz

Excuse me, are you tired? Because you've been running through my mind all night! —@tcoopee

You're single!?! That's unconstitutional. —@MattyP_654

Pogue Sez

Originally, what I asked on Twitter was: "TONIGHT'S BOOK QUESTION: What's the best pickup line you ever heard?"

But when you look over these replies, you have to wonder if these were really the "best" lines. I mean, come on: "You must be a parking ticket, because you've got fine written all over you"? Is that the *best* pickup line, or the *worst*?

On the other hand, maybe I shouldn't judge. You'd be surprised at how many female Twitterers submitted one of these incredible groaners and then added, "This one actually worked on me!"

You dropped your name tag. [Hands over a sugar packet] —@Biha

I know I've seen your face before…or was that just in my dreams? —@pwa1970

[Licking finger and touching her shirt] Let's go back to my place and get you out of those wet things. —@VChrisFurtado

Hello, I'm Mr. Right…someone said you were looking for me. —@coreydahlevent

How much does a polar bear weigh? Enough to break the ice. Hi, I'm… —@KelByrd

Well, here I am. What were your other two wishes? —@drewbiondo

You know what? Your eyes are the same color as my Porsche. —@tokre

I lost my phone number. Can I have yours? —@dhersam

Do you believe in love at first sight, or do I have to walk by again? —@llennon

[Looks at shirt's tag] Yup, just what I thought. Made in heaven. —@kbranch

If I said you had a beautiful body, would you hold it against me? —@roykissel

What's your plan to save the American newspaper?

Product placement. "Senator Blarney, seen here next to a package of Rice-a-Roni, had no comment." —@pumpkinshirt

Stop giving it for free on the web! Your mother was right: No one will buy the cow if the milk is free. —@NickDow

Have every family buy a truckload to insulate the homes they can no longer afford to heat. —@sjacob09

Eliminate the print version. Offer two online subscriptions: monthly/yearly or pay-as-you-go: $.10 to $1.00 per article. —@technic1e

Cheap digital copies, i.e., super-cheap Kindle/iPhone device in newsstands. Put out the daily news on memory cards for said device! —@audioper

Papers should morph into daily magazines covering topics in depth. They can't compete with the Internet, radio, or TV on breaking news. —@tombetz

Re-market them as window cleaning paper—recyclable and earth-friendly. (Well, except for those trees...) —@FrostKL

Autoupdating e-papers with international and ultra-local sections, where citizen and professional reporters collaborate and share. —@JHKramerica

I'm going to read it every day and shop with its advertisers... and hope the bankruptcy court judges do the same. —@SladeWentworth

More schools should use newspapers to teach basic reading, give kids a chance to form the habit. —@pumpkinshirt

Stop selling paper, start selling real journalism: insight, analysis, verification, expertise, professionalism—AKA content. —@smacbuck

Subscribe to at least one. —@ChazEk

Ban toilet paper. —@disser

No way around it: online subscriptions. I hear it works for porn, so why not depictions of reality instead? —@molochhamovis

The newspaper can go the way of the horse & buggy. The important thing is to save American journalism. —@the_gadgetman

More risqué jokes in Mary Worth. —@pumpkinshirt

Charge more for online access than it costs to buy a paper, naturally. Make it less convenient! But then news would be pirated! Whoa! —@kweenie

The same thing that saved British newspapers: Page 3 girls. —@pumpkinshirt

Add the ability to scratch and sniff the pictures. —@nycxaxez

How about a massive puppy-dog breeding program? —@msewall

Why save it when the Internet gives us a vast array of news sources? Blacksmiths were important once too—anyone missing them now? —@BrianNFletcher

The kids in my class sure love the Jumble. Could that go on the front page above the fold? —@pumpkinshirt

The same plan we used to save the American autoworker, coal miner, textile worker and small business owner. —@DrLabRatOry

Cellphone model: Offer readers a free Kindle with a commitment of a two-year subscription. —@fighterchill

Turn newspapers into non-profit multimedia groups funded by endowments. Quit giving away the entire paper online. Synopsis-only is free. —@ChaseClark

Localize, localize, localize, localize, localize, localize, localize, localize, localize, localize, localize, localize, localize! —@davidhitt

Blow up the Internet. Any idea where it's located? —@simonmcd

Let's go back to a tried-but-true method. I'm speaking, of course, of going to war with Spain. —@pumpkinshirt

Instead of charging readers for online content, newspapers should charge ISPs for access. ISPs can add the difference to their cost. —@elliotschimel

Less wire copy, more local features that aren't of the cookie-cutter variety. —@pumpkinshirt

80 pages of Sudoku. —@pjpaul

Genetically engineer parakeets to be 15–20 pounds, forcing everyone to buy bigger cages—to be lined with 7–8 layers of newspaper. —@Gen215

Recession→Depression→Hobos→Cold nights, park benches→Increased demand for newspapers. Especially broadsheets. —@pumpkinshirt

Bake sales! —@reedkavner

Let newspapers fail, wait until people miss them, then watch them rise from the ashes, stronger than ever. —@cbeard765

Redesign them, make the format smaller and make all news be 140 characters or less. —@oscarcortess

Pay high school newspaper advisers more. This has NOTHING to do with the fact that I am a high school newspaper adviser. —@pumpkinshirt

Print it on delicious, edible paper. "Honey, are you done with the sports section? I need to fix dinner." —@noveldoctor

Mail more fish. —@Sandydca

Double-entendre headlines on every page. On a really slow news day, triple entendres. —@pumpkinshirt

Sell it as a package deal with firewood and matches. —@SteveKlinck

Save newspapers? Sure. Put them in plastic Ziploc bags like comic books and store in a cool, dry area. —@scerruti

What should the 11th Commandment be?

Thou shalt not falsely hyperinflate the world's economy based on subprime mortgage–backed securities insured with worthless policies. —@leegoldstein

Watch not any Harrison Ford movie if it be after "The Fugitive." —@pumpkinshirt

Thou shalt not take up 2 parking spaces with 1 car. —@bklein34

Thou shalt not take the last piece of pizza, except that thou first make an insincere offer to let thy friend have it. —@pumpkinshirt

Thou shalt not be so damn picky picky picky. —@msborecki

Covet not thy neighbor's WiFi. —@mpondfield

Thou shalt not confuse cologne for a real shower. —@etreglia

Thou shalt be nice to others, as it is not that difficult. —@CafeChatNoir

Thou shalt not text and drive. Frickin' idiots! —@hughescience

Thou shalt not impose thy values unto others. —@ascottfalk

Thou shalt occasionally give it a rest. —@pumpkinshirt

Compose a haiku that tells your life story.

Sleepless nights for years
Nurse, teach, feed, cry, play, clean, sing
Worth every gray hair —@MetaMommy

Was born analog
Only to toil a lifetime
To die digital —@uncjonny

Beach gal Desert mom
Runner—thinker—doer—Rest
Still a grain of sand —@AuntFun

I'll do it later
Write the story of my life:
Procrastination —@marthasullivan

How hard can I work
To please others and me too
But not go insane? —@musicalrunner

The unexpected
Crept in while I was asleep
So much for my plans —@MissRebeccaF

Working with children
Working with tech contractors
Don't know which is worse —@cstorms

First I wanted toys
Then I wanted wife and kids
Now I want to sleep —@disser

Friends, family, fun, joy:
Gave it all up when I went
To medical school —@KatieMyers

From my early days
I marveled in my genius
Others don't see it —@tgoral

Euro Army brat
A Boston frat, New York flat
And now WA expat —@andremora

Spring, my hair was gold
Then waves of auburn followed
It is pewter now —@passepartout

Living the same day
Always give but not receive—
Life of a mailman —@xbooernsx

Florida native
Escaped to Windy City
Now make snow angels —@hriefs

Traded New England
For a life in the Deep South
Shopping malls the same —@allie912

Slipped out of the womb
Looked around and laughed aloud
Let's stir trouble here! —@whistlingfish

Wanted to do good
Am a lobbyist instead
Mom should not be proud
—@lowercasedts

Air Force Brat who moved
Around a lot but never
Seemed to go anywhere —@dcmacnut

It's hard to fit my
Life's story in seventeen
Syllables; sorry.—@skylineproject

Provide an example of spam from the year 2100.

Hot clones of you in your area want to meet you! Call our Clone Hotline and talk to yourself for free! —@pumpkinshirt

OXYGEN PILLS BY MAIL! Exceed government ration. Sniff up while others' tanks empty. Bank details needed. No COD. —@kvijayraghavan

One-size-fits-all jumpsuit too tight? Lose weight the all-natural way with our synthetic fat-blocking hormones! —@AndyMaslin

Do you owe WorldGov more than $50 billion in back taxes? Get our free holochure that explains how we can help! —@pumpkinshirt

Get fuel for your antique gas-powered vehicle. Click here to find out how. —@BobTV10

Do your part! Cut down on overpopulation. Reduce your sex drive. 25% off your next purchase of Limpgra. —@brentmatsumoto

The warranty on your flying car is expiring in a few weeks. Please tweet XYZCO to extend your warranty! —@satish100

Earn your degree the easy way...through RNA infotransference! Choose from hyperphysics, biocomputing, or hotel management! —@pumpkinshirt

Tired of re-cloning your pet for hundreds of dollars? Make your own clone at home for pennies a day. —@lizardrebel

Best prices on Soylent Green. "No Relatives" guarantee. —@pmahoney87

Earn big bucks from the comfort of your own bio-pod! Our multi-universe-marketing method makes it possible!! —@pumpkinshirt

Greetings, friend. I am a citizen in Moon Sector 7 in need of help with a $7 trillion transaction to a US account. —@kbranch

You've won a new hovercraft. Click HERE!!!!

—@holioli

Best underwater land deals. Low rates for first-time buyers! —@tommatzzie

Beautiful mutants are waiting to meet you! At AlmostPerfekt, we have arranged thousands of human-to-mutant marriages! —@pumpkinshirt

Want to increase the speed of your gene re-sequencing!? Get a free sample of Gene-Tonix today! —@TonightWeParty

My name is Joe Smith. I need your help to get $10 billion out of the U.S. into my safe bank account in Nigeria... —@sheppy

Write a brilliantly gripping first line of a new novel.

Bangalore was muggy the day the package arrived. I let it sit for an hour before opening it, then reluctantly torched the hotel. —@pumpkinshirt

As the cruise ship faded into the moonlight, Harvey clutched the wet gnome and thought: This wouldn't have happened to Kate Winslet. —@A3HourTour

You'd never think buying a goldfish could trigger the Apocalypse, but I guess you really do learn something new every day. —@rkarolius

Although neither had clothes, she had the saddle and he had the jumper cables; both knew things were going to be fine. —@AlanBryner

It wasn't so much that there was a dead hooker in the bathroom that bothered me. —@wilshipley

It was no accident that President Campbell had been elected twice by overwhelming margins. It was a tragedy, but no accident. —@pumpkinshirt

Considering the alternative, leaping out of the 17th-story window didn't seem like such a horrible way to die. —@noveldoctor

Chelsea St. Dean was the kind of cop who gave 100%, leaving no room for romance. Fortunately, Dirk Xander always gave 110%. —@pumpkinshirt

I had never killed anyone before, and I was pretty sure I did it wrong. —@leafwarbler

There are three Kates in this story, and only one of us has a happy ending. —@noveldoctor

Until the fateful day of the hot dog factory tour, Boris had lived a peaceful, sheltered life. —@jdcb42

She chose friends as she chose chocolate and coffee, always seeking the strong and the bittersweet. —@DaliWied

Tim couldn't distinguish between the smells of his burnt toast and smoke from the fighter jet that had just demolished his house. —@jdcb42

I awoke smelling of bamboo and sherry. I knew that once again, my other personalities had thrown a party without me. —@pumpkinshirt

She was another nameless intern, but the congressman hardly cared as he licked his lips & locked the door to his office. —@hriefs

Looking at the receding, orange-and-red glowing remains of Earth, he thought: That was the best-timed abduction ever. —@Flenser1

My name is Ruby. Reuben Ruby. My friends call me Rube. I don't know why. —@pumpkinshirt

In his 40 years on the planet, never had Brad Johanson ever needed anything this badly. Love, money, fame? No, Brad needed a tweezer. —@gametroll

All the superpowers in the world weren't enough to suppress Captain Amazing's first love: the power of dance. —@mikey_c

Are these your teeth, or mine? —@noveldoctor

Whenever Grandpa visited, it was the same routine: Hide the oranges, nail down the drapes, and remember not to mention Alpha Centauri. —@pumpkinshirt

Harold took his coffee, sipped, and fell to the ground, just slowly enough to see the barista's evil grin. —@skylineproject

As it turns out, you can sing all of "American Pie" in the time it takes for a serial killer to decide your fate. —@noveldoctor

It was a dark, black, really super dark, like darker than dark, plus stormy, with hail and lightning and wind and everything, night. —@grousehouse

It was the kind of date that could only have ended as it did: on top of the Louvre, dressed as penguins, and handcuffed to a bomb. —@pumpkinshirt

Summarize a famous movie in 140 characters.

Mouse screws tup on the job. (Fantasia) —@garygoodenough

Please pass the peas. (My Dinner with André) —@jpojman

Tim Curry is a sweet transvestite. It gets weird after that. (Rocky Horror Picture Show) —@Fallen_Woman

The boat sinks. The necklace sinks. Jack sinks. Rose floats. (Titanic) —@vdr

Lather, rinse, repeat: Ad infinitum. (Groundhog Day) —@jschechner

Billionaire dresses up as a bat to protect city from psycho dressed as a clown. (Batman: The Dark Knight) —@crimsong19

Telemarketer won't leave me alone. (Phonebooth) —@justcombs

Guy investigates his wife's murder. Guy investigates his wife's murder. Guy investigates his wife's murder. Guy investig... (Memento) —@rhys

Wait—that one should be "redrum s'efiw setagitsevni yug. redrum s'efiw setagitsevni yug. redrum s'efiw setagitsevni yug. redr..." (Memento) —@umapagan

Plastics? No. Mrs. R? No. Dumb jock hubby? No. Let's get on the bus, Elaine. (The Graduate) —@sparker9

Giant robots come to earth looking for magic Rubik's cube. (Transformers) —@worldexplorer

Paranoid schizophrenic farmer hears voices in Iowa cornfield–turned–baseball field. (Field of Dreams) —@hriefs

Rosebud. Rich kid buys newspaper. Loses money. Makes money. Mistress terrible singer. Builds castle. It's a sled. (Citizen Kane) —@roundtrip

"Top Gun" in cars. (Days of Thunder) —@edprice3

Bond saves Bolivia from expensive water. (Quantum of Solace) —@mvergel

Batman in a metal suit. (Ironman) —@sweyn

Brangelina is born. And some stuff blows up. (Mr. & Mrs. Smith) —@youngmaven14

Young Jedi trained by muppet has hand cut off by father in light-saber accident. (The Empire Strikes Back) —@syphax

It's...about a rug. (The Big Lebowski) —@mattwolfe

Snakes on a plane. (Snakes on a Plane) —@samalolo

Who's had a brush with greatness?

Sat next to Bette Davis in a restaurant in Wyalusing, PA. She fussed at her grandson the whole time for his soup-eating technique. —@bfulford

Twelve-floor elevator ride with Peyton Manning in Minneapolis. I told him, "Have a good game!" (They won. No thanks afterward.) —@mmarion

I literally bumped into Wilt Chamberlain in Philly's 16th St. Station as a kid. He couldn't see me: My eyes were at his belt level. —@meheller

1980, at CNN, walking backwards & talking to someone. Tripped, fell backwards into The Police (Sting, Andy & Stewart). Such gents—caught me! —@ann_walker

I once told Shaq that he had to switch socks with his manager in the green room of Quite Frankly. White socks with black shoes = bad. —@kseifert

I was the manager of Ford of Ocala (FL), and sold John Travolta his new Mustang in 2005. He is a really laid-back guy. —@rlstroud

Chris Rock went into a convenience store. I followed on a dare. He noticed. Staring contest ensued. He likes whole milk. —@beauessai

At a graduation open house for a friend of mine, Neil Armstrong tapped my car while he parallel-parked. —@NCharles

My dad once waited in line for a bathroom in between Henry Kissinger & Rupert Murdoch. —@harrymccracken

I peed at a urinal between Ronald Perelman and Henry Kissinger at the NY Hilton in 1990. —@EricSails

I once used the urinal next to Henry Kissinger at intermission of "Guys & Dolls" on Broadway. Kevin Costner was also in the bathroom! —@nolanshanahan

OMG, I once peed next to Kissinger too. Seriously. —@vidiot_

Stevie Wonder once came into my old music store job at 9:15 (after close) and played for the staff, then bought a few keyboards. —@davidormesher

When he was a kid, my best friend's father was nearly bitten by Malcolm X's sister's dog. —@Khbrookes

Pogue Sez

I can't explain why so many brushes with greatness take place in public restrooms.

I'm even more helpless to explain why so many of these bathroom encounters involve Henry Kissinger.

I'd ask my Twitter followers for their theories, but it'd be too late; this book has already been published.

Went to camp with the kid who sang the B-O-L-O-G-N-A song in TV commercials. You can't buy connections like that. —@pumpkinshirt

I told Cokie Roberts that my mom liked her. In mock annoyance, she said: "EVERYONE'S mom likes me!" —@photo2010

One day a friend & I were standing at 26th & 6th Ave talking about an opera that takes place on a spaceship when Leonard Nimoy walked by. —@gaspsiagore

I met Steven Wright, and was all excited telling him how I was a huge fan. He said: "Thanks." —@stevegarfield

I used to sell hams to Joey McIntyre at Christmas. No lie. I've got the photos to prove it. —@saltzberg

Met Walt Disney on Main Street in Disneyland. I said, "Good evening, Mr. Disney." His reply: "Just call me Walt." —@igenr8

I stepped on Justin Timberlake's foot while getting my pic taken with him before an 'NSync concert when I was 13; I was mortified! —@KatieMyers

I met and spoke with Ross Perot back in the day when he was on the presidential trail—driving rent-a-car around Florida—GREAT GUY! —@warrickmorgan

I met Brent Spiner in an elevator at the "Houston Chronicle." He introduced himself as Ted Danson. I thought it was funny. —@jamest8

Met Bobby Orr at age 11. He handed me an apple core and said, "Here, kid, would you throw this away for me?" I shoulda had it bronzed. —@shayman

I once heard Patrick Stewart use his magisterial, Shakespearean voice...to give directions to the men's room. The best. Directions. Ever. —@pumpkinshirt

I saw Neil Simon buy a magazine. Told story accidentally (to great acclaim) as Neil Diamond buying donut. Was stuck with that lie for years. —@pumpkinshirt

Smokey Robinson & wife came to dinner in '72. They escaped to bathroom before we ate. My mom freaked, suspecting drugs. (They were praying.) —@setlinger

I've been as high off the court as Michael Jordan. Rode in elevator with him at a convention center where he was doing a charity clinic. —@retrophisch

I was a big Mr. Rogers fan as a child in NY. I later moved to Pittsburgh and ran into him on the street. He then really was my neighbor. —@tyoungs

Met Hanna & Barbera. When asked how they picked the order of company name, they chuckled, high-fived, & bellowed, "We flipped a coin!" —@thatadamguy

I was in line behind Serena Williams at grocery. She was buying cake to celebrate Venus becoming No. 1 in rankings. Very nice. —@JohnPHolmes3

Shook Betty Ford's hand in 1976. She didn't let go and didn't let go and Secret Service agent finally swatted my hand away. Ouch! —@curt_m

Liza Minnelli backed into me in an elevator, thinking I was the wall. Leaned on me, unaware, for 12 floors. Shampoo smelled nice, at least. —@grobie

I saw Sean Penn walk past a guy who said, "You're Sean Penn!" Penn said, "I know." —@bucknam

Josh Hartnett in line at US Customs during a computer crash. JH: "Can't we just sign something saying we're not terrorists?" —@StickmaninDC

A friend took William Shatner's order at Starbucks. He was offended she didn't know who he was. She said, "Well, do you know MY name?" —@fourfootflood

Was behind Warren Buffett in a McDonald's drive-thru. He paid and almost drove off without the food. Had to back up and get it! —@JeffFarrar

I had dinner with Alanis Morissette...when she was 11. Really. —@footage

George W. Bush offered me a stick of gum at the 1995 All-Star Game in Arlington, Texas. I declined. —@hatchjt

I once stood next to Jimmy Buffett listening to house band wreck "Cheeseburger in Paradise"—he smiled and walked out. —@iwestminster

I've danced with Bill Gates, flirted with Steve Jobs, and eaten ribs with Michael Dell (but not all at the same time). —@cherimullins

Almost run over by Little Richard in a wheelchair in the ATL airport. ME: "Hey look—Little Richard!" LR: "What's happening, baby?" —@boydcha

Harry Chapin needed a ride to O'Hare after Indy concert to catch plane to see his daughter in hospital. I ended up driving him the 3 hours. —@scotimus

1966, plane delayed, seatmate was Mercer Ellington; Duke and Tony Bennett came from first class to chat. Ecstasy! —@Sandydca

Ran into Michael Moore on Park Av. I asked him about the new "New Yorker" profile of him. He hadn't read it, and asked me to summarize it. —@vidiot_

Met Aerosmith in Houston when they were touring NASA. Steven Tyler asked where the exit door was. I told him to Walk This Way. Ha! —@mgrabois

I drove Plácido Domingo from his hotel to the theatre once for a concert rehearsal. Never driven so carefully in my life... —@Stefaniya

I stayed at someone's house who went to Hugh Jackman's wedding. Does that count? —@Thea_Smith

What's your great idea to improve the cellphone (other than, "better signal coverage")?

Reads my thumbprint so I don't have to keep typing the password to unlock it. —@lningram

Smartphone with customized silent settings for e-mail, text messages, or push services overnight ONLY. Right now, it's just off/on. —@choirguy_

When you can't answer a call, the End button sends a text to the caller saying you're busy and will call them back. —@umapagan

A way to show caller ID info from cell on the TV—'cause I'm that lazy! —@_B_en

Multiple voicemail greetings, one for "in a meeting," "unavail for 2 hours," "on vacation," "in bad service area," etc. —@davidormesher

End the 2-year contract system. Make service, coverage, and features so good no one leaves you! Compete on quality. —@tsmyther

A phone that can check the calendar and turn ringer off during marked meetings. I used to have a Nokia that did this, years ago. —@redya

Cellphone and keychain can locate each other. I usually have one or the other but not both. —@kriskelley63

Clap recognition: You can clap a certain pattern to have your cellphone call you so you can find it. —@juliegomoll

A button that will answer the phone with a message: "Please hold until I can take the call." Useful when you need to leave the room to talk. —@barrybrown

Ringer volume adjusts as a function of ambient noise. —@StevenJCrowley

Built-in system for letting us tag numbers as telemarketers. Shared database means our phones can learn what calls to ignore. —@nicholasbs

Allow phone to be used as wireless keys (for car, home, etc.). Won't have to carry bulky keys anymore. —@vandelizer

Very tiny electromagnetic pulse device that can be used to disable the phones of people having loud, inappropriate conversations. —@megsaint

Built-in breathalyzer keeps you from drunk-dialing your ex! —@wjmcmath

Forget the breathalyzer! How about a sturdy cover that prevents me from butt-dialing my son? —@phylisebanner

MiFi features (phone acts as a WiFi base station for nearby laptops, etc.). —@Proverbio

Enable a remote switch to turn your lost phone from "Vibrate" to "Ring" so you can find it. —@BurtonVapor

A tone that indicates when a cell connection is dropped, so we can avoid talking to someone who isn't there. —@garygoodenough

Feature that automatically disables features as battery life winds down; i.e., check email less frequently, disable 3G, WiFi, etc. —@iluvgs400

Eject button to launch telemarketers out of their cubes (preferably into a nearby lake of raw sewage). —@larrylinn

Small, one-line, black-and-white LCD screen on the back that just displays remaining minutes that month. —@Solimander

A button to simulate a bad signal to end an unwanted call quickly. —@BobJagendorf

Automatic entry of a phone number mentioned in incoming voicemail to the phone's address book. —@MrClement

Childproofing: Key lock, rubber guard, waterproof, etc. —@rylanhamilton

A battery warning maybe an hour BEFORE the phone completely shuts down. —@ljhardin

A switch to turn off the ring for X minutes, duration of a movie/play/concert. So I don't forget to turn it back on. —@BobJagendorf

If you pocket-dial me, I should be able to make your phone beep really loud, because yelling never works. —@ZevEisenberg

Cellphones that work on multiple carriers. No AT&T in one area? Switch to Sprint or Verizon! —@N_Jones

A self-charging battery that converts everyday motion into power, like Citizen's Eco-drive watches. —@xtello

Cellphone battery exchanges at corner store. Turn in a drained battery and get a fully charged one for $1. —@ascottfalk

When parent calls teen and teen doesn't answer, call is autorouted to ring on all phones in teen's vicinity. —@Mainesailor

A small, elegant weekend/evening phone inside a smart-phone... slip it out and leave your email behind! —@barakkassar

Different vibrate patterns, just like different ringtones, for different callers. —@kari_marie

What's the best vanity license plate you've seen (or owned)?

On my Mini Cooper: COMPNS8. —@creativestimuli

IMAVBL (my blue-eyed, blond-haired mom's plates during the '70s). —@tasena

On my van: GROCRY GTR. —@justcombs

On a Honda motorcycle: FENRY. (Get it? Fenry Honda?) —@rponto

OMGLOLWHEE —@Nagler

My mom, a knitter, has LUV2NIT. Some guys were checkin' out her Passat wagon, saying: "Gotta meet that chick: Love To-night!!" —@EdwardWinstead

I saw a VW Bug with the plate: FEATURE (as in,"It's not a bug, it's a feature.") —@relsqui

There's a Smart car here in town with the plate MAXWELL. —@orihoffer

On a Smart car: DETHTRP. —@itdoug

I once knew a guy who drove a hearse as his everyday car. License plate: U NEXT. —@cmerlo441

On a limo: OWNED. —@LibraryBarbara

3M TA3 (you have to see it in your rearview mirror).
—@tllennon

I had GRN GYNT on a full-sized Bronco. Green of course.
—@jschechner

Mine: ADCRE8R. My wife's: GORE1FL. —@BruceTurkel

On a Honda Fit: HISSY. —@garryoakgirl

Friend is avid swimmer. Didn't think it through when she chose LVSWMN. Gets interesting looks from other drivers!
—@bengottesman

NOTCOP on Ford Crown Victoria.
—@yongar

HI OFSER —@rob_kennedy

Red Infiniti with the plate TKT ME. —@Sujeet

I know the man with the New Jersey plate LOL WTF.
—@zigziggityzoo

On a Rolls Royce: THE LIFE. —@shiraabel

Sexy, long-haired blonde in flashy little red convertible with plates: GOTCHA. Even made the men laugh. —@loganberrybooks

I live in Las Vegas. Mine says CYNCYTI (Sin City).
—@dbrewer80221

IM A MAC —@RetroMacCast

WAS HIS —@quakerdan

On my veggie oil truck, the plate is BACON LV. —@smashz

I8CHILE (I'm a chilehead). The plate holder says, "Pain is inevitable / Suffering is optional." —@smashz

For my Prius, I wanted VOLT AIR (because it's the best of all possible cars). Someone else got that, so I went with OHM I GOT. —@garrymargolis

HELLVETICA —@morrick

PE DIEM —@AnneHawley

CMUNIK8 —@LADinLA

I used to have KHAAAAN. —@pumpkinshirt

3XTWINS (Never forgot seeing this one; it evoked such empathy!) —@lindseyb16

On a large pickup truck: RUNUOVR. —@athayworth

Let's hear your idea for a new reality TV show that we'd all want to see.

So You Think You Can Operate? —@Northgoingzaks

Survivor: Compton —@coachpeterson

Quiet Time: Olbermann, Matthews, Limbaugh, King, O'Reilly & Beck are locked in a room. Last one to talk wins. —@hriefs

Who Wants to Be a Milliner?: Kind of like "The Apprentice," but with budding haberdashers. —@MattTheGr8

The Biggest Hacker: Contestants try to hack each other's computers/websites. —@GregGehr

If It's So Easy, YOU Do It!: Have media loudmouths run the country for a week. —@sw0rdfish

So You Think You Can Write? —@la_dandy

Patriot: Contestants must survive in a major U.S. city with only 100% American-made goods. —@techsutra

Prison Makeover: 10 nonviolent convicts take classes and compete for the highest scores. Winner gets early parole. —@jschechner

DIY Plastic Surgery —@GregGehr

Let's hear your idea for a new reality TV show that we'd all want to see.

LifeSwap: Corporate CEOs must trade lives and salaries with their lowest-level laid-off employees for a week.
—@Notsewfast

Are You Smarter Than Miss South Carolina? —@vidiot_

Dodgeball matches between different religious groups. Like Baptists vs. Catholics, or LDS vs. FLDS. —@jcporter1

America's Funniest Rescues —@vidiot_

Top Reality Show: Reality shows compete with each other, eliminate 1 each week (i.e., REALLY eliminate—the show will be canceled). —@girl_out_there

Supply your best "I used to work at…" pun.

I used to work at an orange juice factory, but they canned me because I couldn't concentrate. —@alancshaw

I always wanted to keep the books for a gourmet restaurant, but there's no accounting for taste. —@Krasovin

Pogue Sez

I've actually never heard of an "I used to work at…" pun. For all I know, we invented the category on Twitter one night.

Originally, I didn't intend for a work pun to be one of the questions for the book. I was just sending along a joke I thought was funny. I typed: "TONIGHT'S MEDITATION: I used to work at a blanket factory, but it folded."

But, my Twitter followers piled on, shooting me one similar punny tale after another.

In other words, the question that now appears here, is totally fraudulent. I made it up later to accommodate the punny replies that came in that night.

Hope you'll forgive me.

I thought I had a real future at the greenhouse, but I kept hitting the glass ceiling. —@pumpkinshirt

I used to work in a soda shop, but it fizzled out. —@WPHays

I once dated a woman who worked in a tractor factory, but she dumped me by writing me a "John Deere" letter. —@getsyd

I invested in bottled water, but the market evaporated overnight. —@KillerMango

I used to work for Santa, but he sacked me. —@mac_kal

I used to have a scuba shop, but it went under. —@wjmcmath

I used to work at a pottery factory, but I got fired. —@jadawa

I used to work for a logging company, but our branch got shut down. —@Notsewfast

I wanted to be a doctor, but I never had the patience. —@filmtex

I used to work in a lingerie factory, but I was given a pink slip. —@dudgeoh

I wanted to be a groundskeeper, but I just couldn't cut it. —@dougtoth

I used to be a trapeze artist, but I was let go. —@piercedavid

Rewrite a famous quotation in the style of the half-wits who leave comments on YouTube.

Damn dewd, chill! I only haz 1 life 4 my country! Sry! —@NBB1

1 winz for man, 1 major pwnage for mankind —@bnl771

aN I 4 An I mAkEs A bUnCh Of bLiNd PpL —@Biha

dewd like askin 4 stuff from ur country is like so ghey u shud b doin stuff 4 ur country —@mikemaughmer

2 B R NOT 2B, DAT IZ DE ?., —@cgranier

Let them noobs eat fatcake! BTW- you're really stoopid. Watch my vids!!! SUBSCRIBE TO ME!!! —@loudstone

Dewd, like I don care where u go, but 4 me, I haz liberty or kill me. FTW! —@cgranier

lol!! u r halarios LUVMUTT453..but fear isnt all that scarey u know really the only thing you need to be scared of is being scared —@tjbliss

: I dunno should I live or not thats teh question amirite? —@marshallbock

4score + 7 yrs ago, r 4 fathers cum to USA. —@crimsong19

What's the best prank you've ever witnessed?

Our boss left for 10 weeks. We made her office up to look like someone had been living in it, complete with empty pizza boxes & clothes. —@kpedraja

Back in high school, we completely gift-wrapped a friend's car in newspaper while he was at work. —@applescriptguru

We ordered a baby-food airline meal for a grumpy friend in high school. Told the flight attendant he was recovering from dental surgery. —@alanjcook

We put some bolts in roommate's hubcap. Made a racket when the car moved, but mechanics couldn't find the cause when the car was still! —@BruceTurkel

Southwest flight attendant held up seatback card during safety demo. Colleague had written, "I need a man" on it. Whole plane laughing. —@DavidBThomas

My ninja friends Saran-wrapped my toilet under the seat. Up in the middle of the night to go—and yes, I felt the warmth. —@makinitrite

High school graduation: When receiving diploma, each graduate handed the principal a quarter for the school fund. VERY FULL POCKETS! —@JoseSPiano

Each day for the two months leading up to Halloween, we had a pumpkin delivered to a friend's studio apartment. —@joyengel

On job-applicant interview day, we outfitted our pacifist CEO's office as a hunting lodge—even a bear head on the wall. —@BerniceCramer

HS senior prank: Let 4 greased pigs loose in school numbered 1, 2, 3 and 5. They hunted 4 for hours! —@carechiang

Back in 8th grade, I had two 6th graders convinced—for 2 months—that I only had one arm. —@sheppy

School motto: "Carpe Diem." So, of course, we put goldfish in the sink. —@kirish43

They changed a co-worker's office while he was on vacation. He arrived to an office full of pictures of other people's kids and stuff. —@iNss

High school prank: We stopped traffic in the middle of town by having a tricycle race with hundreds of racers. —@themoderngal

When I was in 7th grade, seniors put a real "For Sale" sign on school lawn (a girl's dad ran a Realty agency). Brokerage got calls for days! —@leleboo

A huge flashing signboard in the front yard of my friends' house when they got home from daughter's wedding: ROOM FOR RENT. —@alytapp

While manager was on vacation, we replaced the door panel of his cubicle with a wall unit. No way for him to get in his office. —@djtucson

I once saw someone take a screenshot of a colleague's PC desktop, then hid all the real icons and used the screenshot as their background. —@Adamweissman

We threw marshmallows on the course while playing golf with some friends. They couldn't find their ball to save their lives! —@samvenable

Kids forked neighbors' yards, instead of toilet-papering. Huge box of forks from Costco stuck all over lawns, bushes, etc. —@HeatherHAL

On a cruise, honeymooners could never find their pics—they looked EVERY night. We'd bought them all—delivered them on last night w/a note. —@Sky1Ron

What's your hidden talent?

I can tie a cherry stem with my tongue. —@jamisloan

I can voluntarily dilate my pupils. —@roderickrussell

I can make the come-hither gesture with my index toes. —@gnarlykitty

My brother & I can blow bubbles of spit off our tongues, one at a time, which then float to the ground. (Cool? Gross? Genetic freaks?) —@UNnouncer

Simultaneously crossing one eye and vibrating the other. —@afox98

Does rolling a quarter off my nose so that it bounces once off a table and lands in a glass count as a talent? —@SusanEJacobsen

I can solve all known sizes of Rubik's cube (including the 7x7). —@MDMstudios

I have a split tongue so I can move it like I'm clapping hands. Most people stare at it wondering WTF? —@ftrc

Playing tennis at level 5 with both hands (two forehands). —@PGerman

I am an echophone: I can repeat your words as fast as you can say them. (My wife says it's annoying as hell.) —@yodaveg

I can take my innie belly button out. —@arielleeve

The ability to make the sound of a kazoo...without USING a kazoo—only teeth + lips! An interesting music alternative to whistling. —@ldgcoach

Bizarrely acute sense of smell. Can distinguish wine varieties, flowers, perfumes, food...when blindfolded, by smell alone. —@ctrly

I can stand on one foot and pull the other foot up to touch my head. (And I was never a cheerleader!) —@Theatergirl62

Writing in cursive completely backwards, like Da Vinci. —@lizardrebel

Ride a bicycle while seated on the handlebars, facing backwards. —@dsrichard

I can light matches with my toes. (Not particularly useful, but it is a hidden talent.) —@hacool

I can wiggle my ears either together or independently. Fun to watch my kids' faces when they try it! —@JJT

I can turn my feet around to point backwards. —@sheppy

What movie cliché drives you nuts?

Whenever aliens attack, they do so only within camera range of an internationally famous landmark. —@gtobias

There's always a parking spot right in front of where the hero is going, no matter how crowded the city. —@k2dbk

Hero gets in fight but his hair and clothes never get messed up.
—@LibraryBarbara

Villain lectures hero, then attempts killing with obtuse/unproven method, leaving hero time to escape. Why not a quick bullet to the head? —@amywebb

Slow, poignant underscoring during brutal battle scenes. (It was once a cool effect but now a boilerplate for any such scene.) —@rponto

Every PC runs "Hollywood OS," which is protected by an unsafe password that when unlocked has ALL the intel you need—on the first screen. —@juarezz

Repetitions (from multiple angles and speeds) of the same explosion! —@rponto

Bomb countdown timers that completely violate the known space-time continuum. —@rponto

Actor sidles up to a bar & simply orders "a beer"...as if any old beer will do. (Sure, here's an O'Doul's.) —@hriefs

When someone hangs up on a character, he hears an immediate DIAL TONE! —@justcombs

The slow, sardonic clap...ugh! —@misterlamb

It's amazing how aliens always seem to land in the USA! —@SilviaSimeonova

Someone is told to turn on the news—and they catch the report from the beginning! —@joshforman

After 1 hr of mayhem, hero saves the world with something he could've done right at the top of the movie. —@rolfje

Pogue Sez

Every now and then, I wished I could have responded to my own nightly question. The movie-cliché question was one of those times.

I'd have happily revealed my own most hated movie cliché: It's when a character is receiving an email message and the text *chirps and scrolls*, one letter at a time, as it appears across the screen.

What is this, 1962? Don't movie producers think that most of the people have seen a real computer by now? Don't they understand that in the real world, text appears, fully formed when you open an e-mail message? It doesn't spit out, left to right, like a teletype machine of old. Not to mention that this kind of chirping and chattering would drive today's cubicle-dwelling office workers absolutely bananas.

Then again, what do I know? Hollywood has a long and profitable history of insulting the sophistication of the American moviegoing public...

What was the biggest waste of money you've ever spent?

I can't believe I'm admitting it, but I bought those $600 breast-enhancing pills when I was 21. (Guess what? They don't work.) —@livnvicariously

Called plumber for leaky faucet. He came out, tightened by hand the hose connecting to the spout, and charged me $200. —@davidpaull

A soymilk maker. My husband dislikes the milk. It either tastes too watery or too beany. (I haven't used the tofu maker, either.) —@joellem

Hubby bought rock tumbler to polish rocks into...shiny rocks. Never used, still in box. Rocks not shiny. —@abaesel

Macworld Expo, 1988. Bought a Felix for $49.95 (or more). Cross between a trackpad and a pinchy joystick. What was I thinking? —@mklprc

Xbox HD-DVD player for one "rented" movie (Batman Begins). That screening cost $185. —@nathanziarek

I paid to see "Ishtar." Ishtar, man. Ishtar. —@OgLuna

I have to confess—I've bought bottled water. —@dabernathy

I bought a few thousand of those Livestrong-style wristbands during the fad, hoping I could sell them! —@EvanFogel

Spending money to get my brother out of jail. He went right back in for stealing bicycles. —@docinaustin

Spent $500 on a multilevel marketing program to sell rare coin replicas. —@gafguru

I got a set of tapes to improve my memory, but couldn't remember where I put them. No idea what they cost! —@SFDoug

Paid almost $300 for tickets to a play. So embarrassed about the price, I couldn't tell my wife, and we didn't go. —@harryhassell

An engagement ring. It sounded like a good idea at the time. —@mgrabois

Gym membership, prepaid 3 yrs in advance. I think I actually went about a dozen times. I called it my "fat tax." —@Stefaniya

Thanks to a $400 winning bid on eBay, I'm the proud owner of a spectacular autographed photo of Muhammad Ali. Yes, it's bogus. —@hriefs

Settling estate, one sibling argued about the distribution—wanted $500 more. Cost us all thousands to have an attorney sort it out. —@Maggie_Dwyer

$1,200 on a vacuum. NO vacuum is worth $1,200. Seriously. —@grousehouse

Lehman Brothers shares, again and again, because "It just can't go any lower, right?" —@eoinmorgan

When I was 14 I spent all my savings ($500) on a 10" telescope that I used once. I lived in Manhattan & could only see a few stars! —@dvdmon

Bought a Man-Groomer to enhance my shirtless bod. If you're pudgy like me, tanning oil & smooth fat just isn't cool at the beach. —@Cazwell

Spent $400 on a pair of shoes guaranteed for life, only to throw them away a few months later for being uncomfortable. —@pkennard

Ashamed to admit this, but I once spent $5,000 on a set of motivational audiotapes to listen to in my sleep. Seriously. —@DrKoob

I bought a brand-new car with my first paycheck, and was completely taken for a ride by the financing. My $8K Neon cost me $24K. —@beth910

Paying to see "Yentl." Nothing else is in the same galaxy of stupid. —@the_interocitor

A $600 pair of jeans. Because I was 24. —@sonnyrd

Caption this photo.

Stop giving the kid Red Bull. —@dramaqueene

The rejected album cover for Nirvana's album, "Nevermind." —@bupperoni

THAT'S the stand-in? All right, whatever! Let's go. The Hobbit, Scene 1, Take 74: Bilbo escapes from Smaug. —@seaslug_of_doom

All-new, Mountain Dew Baby Formula. It's EXTREME! —@llennon

Why do people always leave me lying on these realistic photo backgrounds? It's so disorienting. —@AndyMaslin

Baby micro jetpack available for pre-order NOW!!!! Only $199.99. —@bwalder

Dad's having fun with his invisibility cloak again. —@rkarolius

OMG, Dad was right! She really DOES sunbathe topless! —@free2traipse

Superman: The Early Years. —@jadawa

These natural childbirth techniques are getting ridiculous! —@iangertler

I can also turn an orange inside out without breaking the skin. —@SherryArtBlog

The X-Men's newest mutant, "Baby's Breath," escapes from prison in Hawaii. —@davidastark

Stay dry in your wet suit—new super-absorbent sportswear by Pampers! —@dabernathy

Vladivostok, Russia, MAY09 (AP)—Russian space program begins training of babies for the 25-year-long mission to Alpha Centauri... —@xtello

We finally got the kid out of Shamu's blowhole! —@tmasteve

You act like you never saw a ninja baby before. —@JC_zoracel

YIPES!!! I thought you said it was like bathwater! —@marklwalker

What cool anagram can you make from the letters of your own name?

A drink offer (Erik Dafforn) —@erikdafforn

A friendlier leaf (I do environmental issues. —Arlene Fairfield) —@afairfield

Ninja bee elm (Benjamin Lee) —@_B_en

Biracial den mom (I actually am a mom, so 2 out of 3 isn't bad! —Nanci Lamb Roider) —@nancilr

His weirdo porn tech (Christopher Downie) —@cdownie

Crablike bacon (How exactly does that work out?! —Bianca Blocker) —@Biha

Hardened live wire (David Erin Wheeler) —@Theory

Emailing chief (Michael Feigin) —@mikefeigin

A ninja mocks eel (Melanie Jackson) —@mellowly

Sick ole me (Mike Close) —@mikeclose

Rehearse lots (By the way, I'm a singer! —Theresa Sorel) —@singersorel

Orgasmic able. Hi! :) (Michael Grabois) —@mgrabois

Random hornets (Darren Thomson) —@darrenthomson

Sneaky aliens! (Lisa Ann Eskey) —@eskeymo

Unwashed liar (Darius Whelan) —@dariuswirl

A messy room (I'll admit it, usually quite fitting. — Amy E Roos) —@aymroos

Erstwhile porch rat (Christopher Walter) —@chrisw427

Kinkier Jests Inn (A theme hotel for naughty jesters? —Kristine Jenkins) —@Dreadkiaili

Real king (Erik Lang) —@wlerik

Nude ogling maze (Angel Dominguez) —@angeldominguez

A golden rumor pro (but NOT a "large proud moron." My hubby is a publicist, Leonard Morpurgo.) —@gypsyheart

Barmaids seen (and not heard? —Brenda Massie) —@eBrenda

A prank jet (Janet Park) —@janet1210

Crack murder (Marc Drucker) —@mposki

Darn nice bimbo (Bonnie MacBird) —@macbird

Legs elate him (Leigh Maltese) —@leleboo

Ragweed martyr (And I'm an allergy sufferer. Ironic. —Margaret Dwyer) —@Maggie_Dwyer

No, merely arcane (Carol Anne Meyer) —@meyercarol

A critical poke (A required weapon in all moms' arsenals! —Patricia Locke) —@plocke

Insulted vanilla (I live in a predominantly African-American neighborhood. —Daniel T. Sullivan) —@dansully

Banish a jerk (Rajesh Banik) —@rbanik

Itch strum (I'm a guitar player who suffers from psoriasis; seems quite apt. —Curt Smith) —@curtsmith

Trainable pet (Bettina Pearl) —@bettina27

Eyelid does ogres (Not sure I'm too happy with it... —Diego de los Reyes) —@Dreyesbo

A maniac withered album (Michael Adam Weintraub) —@mweintr

A choice smell (Michael Close) —@mikeclose

He enlightens sewers (Good to know when I'm chased by Jean Valjean. —Lester Sheng-wei Shen) —@lshen

Wannabe lecher (Lawrence Behan) —@LBehan

Jams zap them torque (I sound exciting and unstable. —Matt Joseph Marquez) —@mattmarquez

Unread, sorta. (Much like many of the e-mails in my inbox. —Sadao Turner) —@SadaoTurner

Satan hell prom (Tux accessories include pitchfork, devil horns, and dead flower corsage. —Thomas Parnell) —@TomParnell

Slow goul made this treat (Matthew Douglas Storlie) —@StorlieDawg

Cheese dill sperm (Eeewwww! —Michelle Despres) —@MDespres

Join lust craze! (Justin Lecaroz) —@JC_zoracel

Bean soda gal (Angela Bodas) —@MidnightCrafter

Ramble on wry (Perhaps I am in the right biz, doing PR & writing executive speeches! —Berry Lowman.) —@BerryLowman

Lesbian harms jaw (No comment. —Brian James Walsh) —@bwalsh2

Bed girlfriend? Yes (...OR...) Befriended grisly (Either way...WIN! —Lindsey Friedberg) —@remembrtherain

Unlabeled clans try (It's an important fact. —Leland Scantlebury) —@scantle

Aroma in sour milk (Mark Louis Marion) —@mmarion

Ugh—She has antsy arm! (Martha Shaughnessy) —@SFShag

A nicer you (Eric Noyau) —@noyau

A dwarf fern (Andrew Raff) —@andrewraff

Fatal stock (A. Scott Falk) —@ascottfalk

Voila—I sin! (Anagram of my wife's name, Lisa Vioni.) —@ackrob75

Uh! Grey gonad goo (Gary Goodenough) —@garygoodenough

Ban hubris (Of course, if I did, it'd ruin all manner of fun mythology. —Brian Bush) —@brianbush

Dancer is smelly (Or, if you're Yoda: "Smelly, dancer is!" — Candysse Miller) —@CandysseM

Erotic neck fury (After much deliberation, I've decided it's awesome rather than perverted. —Courtney Ficker) —@CourtF

Grill rat well (Culinary instruction. —Will Gartrell) —@vonschlapper

Enhance ham by propelling (Glenn Michael Bryan Hoppe) —@ghoppe

Unscrew navel camera (Evan Lawrence Marcus) —@evan1t

Him so algebraic (Note: I am a rocket scientist. —Michael Grabois) —@mgrabois

He is forward! (Howard Riefs) —@hriefs

Laugh shine! (Lain Hughes) —@pumpkinshirt

Snorkleman! (Mark Nelson) —@markmnelson

Slow, homesick yak (Or: "My wok likes chaos." —Michael Kosowsky) —@heywhatsthat

A salsa regime (I LOVE salsa...so this made my week. —Melissa Gaare) —@MelissaGaare

Vile salami lunch (Although I love salami! —Michael Sullivan) —@mtsullivan

An evil Dad (That's my dad's name. —David Neal.) —@moksha

Assume Ninja looms (I'm a bit paranoid... —Jason Samuel Simon) —@jasonssimon

Samba errors! (I'm from Brazil, but not keen on samba, so... —Marise Barros) —@marisebarros

Dark scar within (Richard Watkins) —@richwatkins

Hither comedy (Miche Doherty) —@miche

Inherent joy (See, I told my wife that! —John Tierney) —@JJT

Ranch comma rocks! (Sounds like a really hip grammar camp! —Sharon McCormack) —@smccormack

Talks loud urge (And I do voiceovers for a living! Sometimes screaming car commercials...sorry. —Douglas Turkel) —@UNnouncer

Loveable ere nine. (Apparently, people only love me before sunset. —Arielle Eve Bonne) —@arielleeve

Sitcom us (Describes our household of a teenager, 2 dogs, and 2 cats, including a diabetic one aptly named Sugar. —Scot Imus) —@scotimus

Screw Zen, John! (Attack my religion if you wish, but don't misspell my first name! —Jon Schwenzer) —@UncJonny

Strangest eve (Sounds like a goth band. —Steve Stanger) —@tmasteve

Cut sleeved (Trashy, but it's also my new self-proclaimed nickname... —Steve LeDuc) —@steveleduc

Anagram gem, eh? (Megan E. Graham) —@Tymethief

What's your brilliant idea to improve home furnishings or appliances?

Why hasn't there been a radio with a recorder and a timer, as on TVs? —@bruceif

A microwave with a decibel meter. When popcorn stops popping for 10 seconds, it automatically shuts off. —@lizardrebel

Alarm clock with number pad. Faster time input—and when ringing, it could make you enter the first x digits of pi to prove you're awake. —@CourtF

Television that detects eye movement. After a 10-minute non-response, it will slowly decrease volume & brightness until off. —@lizardrebel

Airtight cabinets. Because some of us can't be bothered to reseal packages. —@mattmarquez

Roomba "garages" built into the cabinets at floor level (especially if they also empty & clean the Roomba). —@smashz

A combo freezer/oven that keeps my food frozen and then cooks it, which I can instruct it to do via the Internet. —@homeyjim

A dishwasher/kitchen table with attached plates. Push a button after dinner, and they flip inside. —@hughesviews

Step on/step off faucet control for my kitchen sink—that way I'd actually do the dishes. —@pixelshot

A fridge that uses outside air for cooling food in winter (I live in Canada). —@rye_b

A mug with a built-in thermostatic alarm; alerts drinker when tea is at optimum drinking temperature. I HATE cold tea! —@vonschlapper

I'd love to have a clock radio with a PROGRAMMABLE snooze alarm. Why are they all 4 or 9 minutes? —@monicarooney

Automatic windows that open and turn off the AC when the outside temp cools, and closes them and turns on AC when it gets hot or rainy. —@wlerik

Garbage disposal in fridge...sucks out spoiled leftovers, makes spills easier to clean up. —@leelee77

Exercise equipment that generates electricity. Charge up a few batteries, and you've accomplished something with immediate results. —@UncJonny

Onscreen sports alerts to notify viewer to change TV channel to catch big moment of a game (e.g., bases loaded, close to TD). —@hriefs

Intelligent window blinds with sensors to adjust up/down automatically, based on the direction/amount of sunlight in a room. —@hriefs

Nightstand-coffeemaker combo. Never have to face the day without caffeine first again. —@bklein34

Heated toilet seats. It gets cold in Chicago. —@hriefs

See-through cylindrical fridge with motorized shelves that rotate.

—@MaggieLeVine

Smoke detectors that know the difference between steam and smoke, and can autodial 911. —@yoshionthego

Motorized oven racks to prevent sliding out hot racks by hand. —@MarkRosch

Add a Page button to the TV, which you can press to make the remote beep (so you can find it). —@pattimikula

OLED ceiling wallpaper, so the whole ceiling lights up. —@zigziggityzoo

I want to create a bed pillow that always stays cool to the touch—no more flipping the pillow 50 times while trying to fall asleep! —@heathergee

TV with built-in BitTorrent downloading. (Don't tell the RIAA!) —@kevinpshan

Captain's Chair Toilet, with plush armrests to maximize iPhone game-play time. —@fatsvernon

What's your favorite little-known household hint?

Almond oil or cooking oil removes adhesives very easily! —@tassiadesign

To get ink stains out of shirts, saturate the stain with hair spray, rinse, repeat. —@perryan

To get chewing gum out of kids' hair, work peanut butter into the hair and the gum will slide off. —@perryan

Cook bacon in the oven on a baking sheet lined with foil instead of in a skillet on the stove. When done, fold up foil and discard. —@tomaplomb

Each kid gets a lingerie laundry bag for dirty socks. Wash them in the bag so each kid gets his/her clean socks already separated. —@treasuryman

Cleaning glass/mirrors/windows with newspaper renders a clear, streak-free surface. (Another reason to keep that subscription.) —@mldrabenstott

Easy microwave cleaning: Microwave wet sponge for 1 minute, wipe down. Moisture releases stuck-on grime. Careful with hot sponge! —@MetaMommy

Chill a warm 6-pack by putting it in bucket of ice water with a few tablespoons of salt to lower the freezing point. Then spin cans. —@bonam

Prevent popsicle drips. Cut narrow slit in a small paper plate. Slide pop stick thru slit. —@macpower

WD-40 removes most gummy residues. Isopropyl alcohol does well too. —@QuodAbsurdum

If a dripping shower head is keeping you awake, tie a string to it to let the water travel down. —@thesciencefox

Spill red wine on carpet? Liberally pour AMPLE amount of table salt on spill to draw up the wine, vacuum salt when dry. Almost gone! —@ldgcoach

Pour boiling water through fabric with berry stains on it. The stains disappear. —@megsaint

Keep a stash of trash bags in the bottom of the trash barrel. When you go to swap out the old bag, a new one is ready!—@DJRDJR

Put an onion in the freezer for 5 minutes or so before cutting it up to avoid tears and watery eyes. —@joshmparks

Wash fruits and veggies by wetting them and scrubbing with baking soda. Rinse well. (Keep a box out—dip wet fingers in it.) —@bobcat1347

Club soda neutralizes grape stains (wine) if applied quickly. —@AlanEdinger

For snooping into sealed envelopes: Freeze for a few hours, then slide a thin knife under the flap. The envelope can then be resealed. —@etreglia

Use a piece of white bread to pick up shards of broken glass. —@ChrisNBC13HD

If you want to see whether an egg is boiled or not, spin it. If it wobbles, it's raw. If it spins fast, it's boiled. —@superbalanced

Combat frizzy hair by rubbing a static cling sheet on your head. —@sgoodin

Store fresh mushrooms so they can vent. I use brown paper sandwich bags. A sealed container will render them slimy and useless. —@jdfinch

Ants can't walk across chalk lines. —@pennifred

Need a quick approximate measurement? A standard dollar bill is about 5 inches long. —@irishhitman67

Questionable eggs? Put in a bowl of cool salted water. If they float, they've gone bad. —@mnthomas

A cup of white vinegar in laundry removes odor from workout clothes/sports equipment. —@mnthomas

To remove candle wax from fabrics or carpets, place a paper towel over the area, then iron with a clothes iron. Wax will melt onto towel. —@tassoula

For stains on silk: Apply white toothpaste and let dry for a day, then hand wash with soap. (Copyright: My mom) —@iNss

Organize your take-out menus by delivery speed. —@mauradeedy

To recover hardened brown sugar: Put it in a shallow bowl and place a wet washcloth over it overnight. When you wake up, it will be fluffy! —@tassoula

Chopping up a lemon and putting it down your garbage disposal helps disinfect your disposal and makes it smell nice! —@HillaryMoney

To clean permanent stains from inside those coffee mugs, fill 3/4 full with water and 1/4 with Clorox bleach. Wait 30 mins & wash. —@Sandydca

To unscrew a broken lightbulb, cut a potato in half, press the flat side against the socket, and turn. Works great. —@superbalanced

Reheat pizza on the stove in a dry frying pan. Crisp crust! —@Jeff_Bernstein

Put a damp cloth in the dryer with clothes to remove wrinkles. Easier than ironing. —@keikosan

Rub your hands on your stainless steel sink to remove the smell after chopping onion or garlic. —@fwalrod

Food burned onto the bottom of a pot? Boil baking soda in the pot and it will magically lift away. —@bdc99

Rub a slice of onion on small burns. It takes away the pain and they won't blister. It works. —@laffingbird

Finish a familiar saying in a new way.

A journey of a thousand miles begins...with a comfort stop and a return home to retrieve forgotten belongings. —@joschlegel

Revenge is a dish best served on live national television. —@pumpkinshirt

Sticks and stones may break my bones very, very badly. —@Pneumophil

Give a man a fish, and he'll eat for a day; teach a man to fish, and he'll be gone every Saturday for life! —@KillerMango

If the shoe fits...buy two. —@livnvicariously

Hell hath no fury like the owner of the original model when the 2.0 comes out. —@pumpkinshirt

A journey of a thousand miles...can earn you double flyer points if you fly this weekend! —@drudra

A penny for your thoughts—$5 if you keep them to yourself. —@sjacob09

Cleanliness is next to impossible! —@MatthewDooley

A penny saved is...all I have in my 401k. —@justcombs

A rose by any other name would just confuse everyone.
—@brianwolven

Let a smile be your umbrella...and you'll end up with a mouth full of rain. —@channel001

When the moon hits your eye like a big pizza pie, that's both astronomically and ophthalmologically catastrophic.
—@MikeTRose

Find a penny, pick it up, and all day long you'll have...a penny.
—@UNnouncer

Where there's a will, there's a long-lost relative fighting in probate court. —@lizardrebel

Sticks and stones may break my bones...but words can cause irreparable psychological damage. —@sjacob09

People who live in glass houses...shouldn't run around naked.
—@stephanie2967

A penny saved is...most likely in a jar on top of your dresser.
—@pumpkinshirt

He who is without sin...has a lot of catching up to do.
—@pennymonger

What's the cutest thing you've ever heard a kid say?

Four-year-old, asked what mommies are made of: "Sugar and spice and everything nice." Daddies? "Salad and bacon."
—@m0nkeyh0use

Evening bike ride with my 4-year-old, lots of crickets chirping. He put a finger to his lips & says: "Shhh, I hear little screaming!"
—@UNnouncer

Teacher: "What is California's state gem?" Student: "24-hour fitness?" —@enjoyva

Daughter, almost 3 years old, "helped" me take interior house measurements. Me: "How long is the hallway?" Daughter: "Half past six!" —@jsarsero

Kid saw my aunt's hairdo (remember when they were called that?) and asked if it was defrosted. —@smashz

"My allergy test says I'm allergic to dogs and cats. So you need to buy me a pony!" —@hughesviews

"Look, Mommy, he's dot-to-dot!!" (describing a freckled man)
—@pamheld

"I put a pyramid at the end of my sentence." —@hughesviews

"Mommy said you need to clean your room, and I will help you." "Mommy, can you help me clean my room without me?" —@wlerik

My youngest, age 5, shooed us all out of the bathroom & closed the door, exclaiming, "I need my private-seat!" —@Anneliseh

Four-year-old at the movies, puts on 3D glasses: "Nothing looks 3D yet!" —@renelae

2.5-year-old, loudly in store: "I'm high, Mommy!" "Excuse me?" "I'm high!" [raises hands over head] "Oh, you're tall?" "Yes. I'm high!" —@TrishFreshwater

When I was 5, I asked my aunt on her birthday how old she was. She answered 29. I yelled, "You're 29 and you're still livin'?" —@maineroots

Four-year-old sister responds to the hairdresser's question of how she'd like her hair cut: "I want it cut longer, please." —@andyn

Years ago, my parents took me to McDonald's. I asked for a "sweet steak." Huh? (At the time, McD's on TV was promoting a sweepstakes.) —@margaretmontet

My 2-yr-old, seeing his first convertible: "They forgot to close the lid." —@nickfruhling

A very little friend (2 years old?) seeing hail for the first time: "Look! Baby ice!" —@susanbdot

My 5-year-old daughter loved to swim in her "baby suit." —@mrsseigleman

"If I don't win first prize, maybe I'll win the constellation prize." —@hughesviews

Four-year-old counts down to something: "10-9-8-7-6-5-4-3-2-1...blacksocks!" —@jarmstrong58

My almost-4-year-old son was looking out the window at the rain and said, "It's a cold, dark world out there."

—@benjamin_gray

Student asks on trip to aquarium: "If octopuses have 3 hearts, does that mean they can love 3 persons?" —@matt11281x

"Jake, why do we all hold hands when we cross the street?" Jake (age 5): "So we can all die together." —@NeilRaden

My 5-yr-old son, after hearing that all we wanted when he was born was a healthy baby: "Well, you should have asked for superpowers too!" —@webpoobah

[Singing] "Frosty the snowman—had two eyes made out of cold." —@hughesviews

Last week, my 7-year-old's lunchroom chat with his friends was about switching from Comcast to Verizon... —@reganref

Me: "I need you to clean up your mess on the floor." 4-year-old: "Blah, blah, blah, YOUR needs!" —@victoryfarm

Airport X-ray, kid reading pictograms of banned items: "No guns, no knives, no pirates." (Last item, corrosive items, rep. by skull & crossbones) —@mellowly

To 7-year-old scarfing down potato chips: "Are you going to eat that whole bag?" "No, I don't eat bags!" —@rllewis

My daughter once asked me why I had to go to work in order to make money. She asked, "Why not just go to the ATM, like mommy does?" —@johntyleski

Child confidently pointing one by one at letters on fast-food trash can door: "T-H-A-N-K Y-O-U... Garbage!" —@ryaninc

Three-year-old asks for food at a wedding buffet. Older relative: "What's the magic word?" Child: "Abracadabra?" —@TeresaKopec

"It's getting late; it's almost bednight!" —@hughesviews

In disciplining then–4-year-old son. Me: "Who's in charge in this house?" Him: "God." —@ColleenHawk

At an ice-cream stand with 5-yr-old nephew, he asks for mustachio-flavored ice cream. —@LaurieP

Niece talking about a classmate: "She said she had petunia. I think she meant to say menonia." —@ginabean

Four-year-old daughter, when told to pick up extra toys on her way to room with 1 toy while a pile sits next to her: "I'm not an octopus!" —@JamesDevlin

"What do you want to be when you grow up?" "A petinarian." —@Mallatabuck

Daughter: "Where are my stuffed moose's boogers?" Father: "Where are your manners?" Daughter: "Where are its boogers, please?" —@jonedm

My mom to my 6-year-old nephew: "Once, when I was your age..." My nephew's response: "Well, once when I was YOUR age..." —@mikemaughmer

We used to have to cross a toll bridge regularly. My just-learning-to-read son asked, "Mommy, why do we always have to pay the troll?" —@Bonnie_Rodgers

"When I grow up, I want to be an action figure." —@CharJTF

Friend's 3-year old daughter, Nishka, naming vowels: "A for apple, E for elephant, I for iPod..." —@vdalal

"I hope heaven has cable." —@hughesviews

"Mommy has a Honda. Daddy has a Toyota. I like Daddy's car better, because it has Yoda in it." —@badassdadblog

"I think she got a crack in her brain and all of her think leaked out." —@larsperk

Eighteen-month-old yelling, "Hello...Hello?" into a TV speaker, thinking someone is in there. (Hey—the phone speaks back, right?) —@ramganesh

During a visit to the vet for our injured dog, daughter (age 7) gets on pet scale and asks, "How much do I weigh in dog years?" —@BrianAlpert

My daughter, looking for her bathing suit: "Have you seen my zucchini bottoms?" —@geaux_girl

I told my then–4-year-old son that you don't say "Yes, sir" to women. Me: "Do you know what you should say?" Him: "Yessirree?" —@SharonZardetto

Three-year-old: "Mommy, when I grow big and you grow small, I'll do your hair for you, OK?" —@Stefaniya

My daughter: "Why are we having a war on a rock?" —@Trick_or_tweet

My 4-year-old son asked, "When are we going to the marijuana club?" (He meant "Awanas Club." Oh, the looks I got!) —@rockybg

Minivan cut us off in a parking lot. Friend's 4-year-old says: "Damn bimbos!" (I blew coffee all over the dashboard.) —@jasonmuelver

Make up a new Internet rumor that sounds just real enough to catch on.

One in 10 babies was born with a birth defect or autism in 2009 due to mother's stress about economy during pregnancy, doctors say. —@AndyMaslin

The swine flu virus has mutated again, is more deadly than previously thought, AND is passed on toilet seats. —@technic1e

Congress trying to pass law to tax for Internet access. Call your congressman! —@AndyMaslin

New study determines Southern mosquitoes bite less frequently, are "politer" than their Northern counterparts. —@VenetianBlond

The US gov't financed the bailout with the deeds to National Parks as collateral to China. —@endeavor7

Pepsi introduced a can featuring the Koran and promoting the Muslim faith! Boycott Pepsi this summer! —@AndyMaslin

Rick Astley actually lip-synched "Never Gonna Give You Up" over the voice of his 50-year-old, 250-pound uncle. —@bnewmann

Rush Limbaugh has been secretly voting Democrat all these years! —@SullivanDavid

Having a heart attack? Call the hospital and they can use the defibrillator on you over the phone. Just hold your phone to your chest! —@hornsolo

Increase in home robberies attributed to Google Map street views. —@EShahan

USDA has uncovered traces of bird dung in the glue on postage stamps. DO NOT lick "Forever" stamps! —@hughesviews

AOL is officially discontinuing its dial-up service. Also changing the "You've got mail" voice. —@Solimander

David Pogue's Twitter account is actually run by a 10-year-old girl in the Philippines. —@rolando

Add 1 word to a famous name; define.

Stephen King Kong: The 800-pound gorilla of American fiction —@pumpkinshirt

Boy George Bush: Culture country club —@johnmarkharris

Mary Tyler Dinty Moore: "Well, it's stew, girl, and you should know it…" —@pumpkinshirt

Deep Sore Throat: Tattletale with swollen glands —@sjacob09

Peeping Tom Hanks: The creepiest, nicest guy in Hollywood —@sherrynoik

Steppenwolf Poo: Something to be avoided —@sjacob09

John Wayne's World: Schwing, now, pilgrim —@pumpkinshirt

Jack Daniel Pinkwater: The drunkest commentator on National Public Radio since 1974 —@pumpkinshirt

White House, MD: A government that can solve every problem within an hour…and do it again next week —@lizardrebel

Fat Albert Einstein: Hey, hey, oy vey! —@ChuckEye

Hank Aaron Copland: Composer of Appalachian Swing —@rponto

Plaster of Paris Hilton: How socialites get cast —@maryegilmore

Jimmy Stewart Scott: Merry Christmas, Bedford Falls! Booyah!! —@MyCatIsOnFire

Tom Tom Hanks: "Mama always told me to turn LEFT in 300 yards." —@EvanFogel

Pogue Sez

You couldn't help but notice: As they responded to my nightly book questions, a lot of my Twitter followers made fun of Republicans. Out of 25,000 submissions, 91 made fun of Bush, Cheney, and Palin. Only 3 responses made fun of the Obama administration.

Does that mean that there are more Democrats than Republicans on Twitter? Or that Democrats are more likely to contribute to a book? Or just that Bush-Cheney, having left office, are now officially considered ripe for teasing?

Whatever your political persuasion, one thing's for sure: the presence of Bush-Cheney jokes in this book is not the result of some editorial predisposition. It's just that very few people submitted jokes that made fun of the other side.

Rush Hour Limbaugh: A huge traffic jam for no apparent reason —@itfrombit

Ansel Adams Family: A clan of slightly creepy, kooky, mysterious, spooky, altogether ooky professional photographers —@rondavison, @char_anderson

Pee-Wee Herman Wouk: Acclaimed author of "The Winds of War, But What Am I?" —@pumpkinshirt

Blair Waldorf Salad: What was served at Gossip Girl graduation lunch —@AnnieLaG

Marco Polo Marco: One man's journey of self-discovery —@iDoogie

Michael Jackson Jordan: NBA star with a trademark moonwalk shot. Eee-hee! —@JimFl

Don Julio Quixote: Makes you tilt LIKE a windmill, not at one —@drewallstar

SpongeBob Newhart Squarepants: Absorbent shrink with wacky friends —@pmahoney87

George Busch Beer: The misunderestimated king of beers —@SubtleOddity

Condoleeza Rice Cakes: A bland government official who leaves you with a mouthful of unidentifiable nothings —@etreglia

Brian Williams-Sonoma: Nightly cooking news —@CRA1G

Tiger Bretton Woods: Conquered the international finance game at an early age —@pumpkinshirt

Mad Max and Ruby: Anthropomorphic young bunnies in a post-apocalyptic world —@dudgeoh

Dirty Harry Potter: I know what you're thinking. Did I cast six spells or only five? —@pumpkinshirt

Sherwin Roger Williams: Paint used exclusively in Rhode Island —@ddauplaise

Tom Brady Bunch: A passel of skilled quarterbacks from the same family —@AndyMaslin

James Oscar Madison: The fourth, and messiest, president

—@JimWay

Ground Chuck Norris: You don't eat this meat; it infiltrates your stomach, then roundhouse-kicks its way out

—@thebandnork

Shirley Temple Emanu-El: A non-alcoholic drink for kiddush

—@markrhammer

Raggedy Andy Rooney: You know what really bothers me? When my button eye falls off and has to be re-sewn...

—@pumpkinshirt

Invent a restaurant concept that's likely to fail.

Luftwaffles (World War II–themed waffle bar)
—@the_dumb_waiter

Squiddables —@suzanway

Krispy Krill —@pumpkinshirt

Steak and a Haircut
(Get a haircut while you eat!)
—@superbalanced

Squat and Gobble —@sarahbeery

Purina Moist & Meaty Ale House —@SusanEJacobsen

Soupsicles —@garygoodenough

Liver King —@PrompterBob

Sam & Ella's Chicken Grill —@SlauBeSharp

Escargot2Go —@whistlingfish

It's a Delicacy! (monkey brains, crickets, grubs, etc.)
—@lizardrebel

George F. Will's Soul Food Bistro —@pumpkinshirt

Steak in a Cake —@JimFl

Mr. Broccoli —@Enclarity

Charred Beyond Recognition —@pumpkinshirt

Sauerkraut's —@angeldominguez

Comrade Jerk (Russian/Jamaican) —@kentkb

The Nood (a clothing-optional noodle bar) —@pumpkinshirt

Fear Factor Palace —@aramc

Taste O' Des Moines —@pumpkinshirt

International House of Leftovers —@smccormack

Contagious (food trends from around the world) —@KillerMango

Sushi You! (the do-it-yourself sushi experience) —@EvanDGotlib

Poup Scoop (nothing but Dijon-flavored ice cream) —@alytapp

Le Taco Belle (the finest French/Mexican fusion) —@Wildecat101

Soy Vay (a kosher restauraunt for the lactose intolerant) —@kevinpshan

The Sushi Grill —@kvetchguru

Chez Roughage —@pumpkinshirt

Moshe's Vegan Barbecue! —@Stefaniya

Kill It 'n' Grill It (Slaughter your own dinner before dining) —@aaronmiller

Innards to Go! —@hannnahkl

Unfriendly's —@kevinpshan

FishBlender Smoothies —@noveldoctor

Donner Party Lodge —@davidvnewman

Spam-u-topia —@Doublelattemama

Drive-Thru Japanese Hibachi —@MrPromee

Haggis King —@Notsewfast

The Empty Dumpling —@danblondell

Vegan Hooters —@Jimbrez

Tell us the story of your tattoo.

My student trumpet player got a tattooed excerpt of Stravinsky's "Soldier's Tale." In sheet-music style. That's dedication! —@rponto

My son has a fractional representation of the golden ratio: 1 + 1 / 1 + 1 / 1 + 1... around bicep. (He's a math major.) —@ChicagoDiane

I went to a dyslexic tattoo artist & got an ANGLE. 3 weeks later, I got it covered up with a solid angel. —@CharPrincessa

In '96, my Mac was so slow with Photoshop that I worked on an ankle tattoo to kill time. Then I got a new Mac—have only 1/2 a tattoo! —@MyMukki

My Twitter picture is the robot I have on my right arm: a reminder not to be a fraud, mechanical, or conform to society without questioning. —@Spinzley

Lost and depressed when I dropped out of school. I got stars on my foot to remind me to keep looking up whenever I'm down. —@ndpittman

Secured a custom major at CCSU (digital humanities). Got a USB tattoo on my right inner wrist, near my primary human-interface device. —@AlxJrvs

Hard to explain, but it goes down to a strange "Lord of the Rings"/Hobbit phase. —@mariadcgomez

The only way they'd break a $100 bill for me was if I paid $20 for this Pokémon on my shoulder... —@xtello

I love NYC as if I were a native, but I live in Colorado. I have the skyline of Manhattan tattooed on my arm to remind me to keep striving. —@DrewJazz

My dad has 2 on his butt, of devils shoveling coal you know where. Army days, apparently...I don't ask. —@marchdigital

At the U. of Wyoming, I got tired of coeds asking, "Who's Linda?" So I had it covered by a full-circumference cobra. —@jeffleroydavis

Caption this photo.

Dude! I think it's some kind of stargate! —@rationalorder

You kids have it so easy. Back in my day, we foraged and stored nuts for the winter! —@CathleenRitt

NYTimes Special Report: HOUSING CRISIS HITS SQUIRREL COMMUNITY HARD. —@r1goff

Hey, guys!! I can hear the ocean in here! —@bneuman

Stay right here. I'll be back in a JIF. —@JMWander

Honey, does this jar make me look fat? —@MichaelRubin

I hate the way it sticks to the roof of your head. —@bojack54

Sorry, I thought you said *I* was chunky. —@Stilgar702

Squirrels: Good at eating nuts—bad at hide-and-seek. —@AATBenjamin

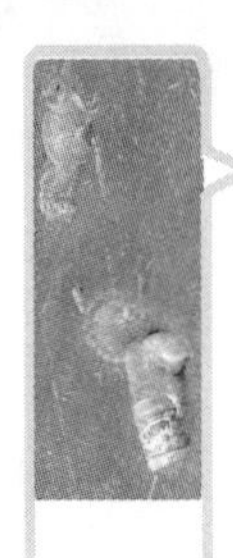

Caption this photo.

One solution for two problems: (1) Squirrel infestation. (2) Inventories of salmonella-tainted peanut butter. —@ikepigott

Science's earliest clue in the puzzling evolution of the Long-Tongued Gray Squirrel. —@ruralbroadband

Funny...he never asks for seconds of Skippy at home... —@bricktop01

You gonna finish that? —@djmsalem

What's your strangest habit?

I keep the temp in my car set at odd numbers that somehow involve 3. I'm also a leg jiggler. —@SueMarks

My husband, after washing dishes, always has to leave one dirty dish in the sink. —@AnnieLaG

I drink stuff out of measuring cups as opposed to real cups. —@yoshionthego

I must have everything in alphabetical order: movies, CDs, books, food in pantry, my wife's perfume, etc. —@betaboy78

I draw on my palm or thigh words that I hear while watching TV and movies. I don't know why, maybe I have a cursive fetish. —@Juspeter

I HAVE to have all paper money in my wallet/pocket all facing the same direction in smallest to largest denominations, ALWAYS. —@Wolfwings2

When asked about a date in the future, I look at my bare wrist. I haven't worn a watch in 20 years—and that watch didn't track dates. —@jadawa

Sometimes, just before sitting on the couch, I make a full turn to the right (360). Go figure... —@angeldominguez

Biting off all the kernels on popcorn & eating the soft balls after I'm done. (I trust you will keep this in strictest confidence.) —@indiereads

When I buy something at the store, I always pick the item in the back of the shelf, never the front. —@BlueGromit

I count the number of steps as I walk up and down stairs. Wait, did you mean strange AND interesting? —@megsaint

Organizing store shelves while I'm out shopping. Annoys my friends to no end. —@aquinonez

I pretend I'm using The Force when going through automatic doors. (Shameful, but true.) —@joshgans

I hold my breath every time I go over a bridge. I have a constant fear of earthquake/collapse. I live in KY. —@mslisacakes

I don't like to eat things in odd numbers. 3 cookies? Gotta be 2 or 4. Things like to be eaten in pairs, didn't you know? —@christopherbmac

I always double-knock with my right fist the fuselage of every airplane I board. It's my weird way of making sure the plane is sound. —@mweintr

Halfway through washing the dishes, I have to dry my hands because they feel too wet. Then I carry on with the dishes. —@zoara

Brushing my teeth with three different toothbrushes every time I brush. No, not obsessive-compulsive...I think. —@pinlux

I sort my M&M's by color before eating them. Need more greens. —@jrb1165

Making sure the light switches at the top and bottom of the staircase are facing the same direction.
—@gizmosachin

I only wash my hands with really cold water. They don't feel clean if I wash with warm or hot water.
—@Theatergirl62

My wife gets totally congested after meals. We think she's allergic to all food. :)
—@sheppy

The alarm I wake to must be set to an even number.
—@ElWray

After washing my hands in a public bathroom, I only use 1, 3, 7, or 12 paper towels—don't know why.
—@johnmarkharris

Pogue Sez

This question produced an enormous outpouring of responses (of which only a sampling appears here). From this, we can conclude that almost everyone has one weird little compulsion or another.

What amazed me, though, was the number of people who have the *same* weird habits. Three people all reported—completely independently—that they double-knock the outside of a plane as they board.

If you can believe it, nine people all said that they sneeze three times when they're finished eating. Not twice, not four times—three times. What is *that* about?

Then there are all those numerological rules, the hand-washing oddities, the object-organizing compulsions... it's kind of comforting, actually, to know that everybody is pretty much equally nuts.

I grip the steering wheel of my car tighter when I cross a shadow across the road. I wish I were joking. —@rondavison

My friend refuses to drink out of a glass unless it has 5 ice cubes in it. —@GaleMcCarron

Eating salted peanuts, I grab a handful, shake my hand 3 times, pop a few in my mouth, repeat 3 shakes, eat more, etc. —@DoctorTom

I never put down a book on page 13 or 87, 113 or 187, etc. I've discovered, though, that it makes no difference to my life. —@dkahans

My brother sneezes 3x whenever he's finished eating. Helpful, actually—we all know he's done. —@Mainesailor

I find it almost impossible to speak a person's given name, must use title or nickname. —@chemrat

If I snap my fingers, I HAVE to do it an odd number of times, or I feel all weird. —@jst79

I used to never eat ends. That rule applied to an amazingly large variety of foods. —@etachoir

If one of my hands gets wet, I can't dry it without first rubbing both hands together so they are equally wet. —@dldnh

I use Proper English on the Internet and I'm 13 years old. —@TheAngusBurger

Men: What do you wish women knew about you?

Men are astounded by the number of lotions, potions, & creams women have, & wish they could tell women that none of it seems to work. —@passepartout

Women need to know that men need to be respected, not coddled or denigrated. A kind & soft woman wins a hardhearted man every time. —@datakcy

A good sports matchup always presents a conflict of interest in any situation—i.e., weddings, childbirth, etc. —@AndyTheGiant

Sometimes when we say we're thinking nothing, we're really thinking nothing.
—@ngrach

For men, no means NO and yes means YES. There is no wiggle room in those two words. —@jjanssen15

I wish women knew that we can't read their minds, and it's not because we don't try. —@mattstafford

Just because we don't remember things doesn't mean we don't listen. It just means we don't remember! —@BerryLowman

I wish women knew that all we want is beer and sex. —@dremin

If you say one thing and mean the other, we will always listen to the one thing. Deal with it—it's not changing. —@Khoji

That we (men) aren't nearly so smart as they give us credit for. That means we aren't as devious or as malicious either. —@emd

If I can't tolerate you when you're sober, drinking isn't going to magically solve the issue. —@esjWBRU

Saying, "Always be honest" and then asking, "Do I look fat?" or "Do you think she's pretty?" are mutually exclusive. (And terrifying.) —@jonnydeco

Like vents for a hard disk drive, holes in men's underwear are strategically placed for ventilation purposes. —@DaveSachioMori

The "just a minute" a woman takes to get ready is roughly equal to the "just a minute" left in a televised sporting event we watch. —@dhersam

Leaving the toilet seat up is determined by our DNA. Just ask Watson and Crick. —@hriefs

I wish all the women who have said "I have nothing to wear" would tell whoever owns the clothes in their closet to come get them. —@whistlingfish

Women: What do you wish men knew about you?

Wife sick in morning
Husband heeds warning
Wife sick at night
No cuddling in sight —@Doit2it

Sometimes when we want a backrub, we want ONLY a back-rub!! —@cstorms

That women pay much more attention to men's body language than they could ever imagine. We see and analyze all. —@InaHurryMama

As a woman, I don't want the men in my life to solve my problems. Just listen. Do not, do not pontificate. —@Rox_N_Stone

I wish that 90% of men could understand that women have a sense of color—there's more out there than dark colors & khaki. —@kweenie

I wish tall men knew they had to trim their nose hair. It's distracting and tends to catch mucus. Ewww. —@CharPrincessa

You'll never completely please a woman, but we'll love you for the effort as much as the result. —@auldfatbroad

Women really hate when guys talk to their breasts. (Wish I had $1 for every time this has happened to me.) —@mathheadinc

Women also golf, hunt, fish, hike, work on their car, build things, etc. Join them or get out of the way. Stop feeling intimidated. —@LorieMarchant

When we need to complain and rant about something, we don't need the guy to fix it. Just listen and sympathize—let us rant. —@abaesel

We want you to know we WILL get old, and we are afraid you won't love us anymore when we do. —@bobcat1347

We women don't want to do it ALL. Pick up your own dang laundry. —@MetaMommy

We want men to know the heart does not compute. —@macbird

We want men to act like our best friends—listen, get emotional, etc.—but when you actually do, we're turned off. —@tesspantaz

We all want to be validated. "Yes, honey, I hear you and I understand." That's all! —@ShanKSizemore

The only way we know you're listening to us is if you are looking at us. —@myfaka1908

We secretly wish men had to wear high heels too. —@BetsC

When a fight gets to the point where you forgot what the fight was about, just kiss her!!! —@plumapluma

Wish they knew that PMS is real and not just some excuse! —@becavale

That even though we have 30 pairs of shoes, we will always need another pair. —@Ssaskie

We wish men were a bit more intuitive and better at picking up on details and subtle nuances. —@pixelesh

We want men to understand that sometimes even WE don't know what we are thinking. —@keikosan

This is to the men who feel deprived: Hugging & kissing & doing the dishes will = more sex. (I'm a female psychotherapist.) —@cynthiamckenna

Men need to know that women have to take off their makeup eventually and even then, they're still beautiful. —@yulimar

We use our peripheral vision to gauge your reaction to any hottie. We see everything, so don't drool or be a fool! —@Pawskya

We want men to read our minds (there, I said it). We mean the opposite of what we say a lot of the time and expect men to get that. ;-) —@jazcan

How many men does it take to replace a roll of toilet paper? We don't know. It has never been done. —@Sandydca

Make up a likely headline from the year 2012.

'Save United States' Cause Raises Billions for the Floundering Third-World Country; India Steps Up —@SurHorse

Studies Show Obesity Linked to Overeating —@brieana

Arnold Schwarzenegger Retires from Politics, Returns to Acting —@AndyMaslin

Amazon Jumps on Newspaper Nostalgia Bandwagon with New Broadsheet-sized Kindle —@rponto

Erratic Bill O'Reilly Embroiled in On-Air Feud with Himself —@pumpkinshirt

Dec. 31, 2012: "Eat It, Mayan Prophecy, We're Still Here!" Jan. 1, 2013: "World is Ending, Thanks to Yesterday's Headline" —@G4Flobot

Fresh Air Hits a Yearly High of $2.99/Can —@EvanFogel

In Desperate Bid to Boost Circulation, NYT Adds Maureen Dowd Comic Strip —@pumpkinshirt

US Government Grants Fiat $7 Billion Bailout —@justcombs

New Poll: Americans Still Confuse Memorial Day, Veterans Day Holidays —@jdonels

Fox to Premiere Reality Show That Follows Producers as They Attempt to Come Up with New Ideas for Reality Shows
—@pumpkinshirt

Rabbit Flu Outbreak: Is It Time to Panic? —@i_Girl

Pogue Sez

A lot of the questions posed to the Twitterverse for this book generated duplicate responses many times over. When I asked, "What spoken phrase drives you crazy?", dozens of people responded with the phrase "I could care less." When I asked for the best bumper sticker anyone had ever seen, I got a lot of, "My kid can beat up your honor roll kid."

And when I asked for headlines from the year 2012, I got scores of "Obama re-elected" jokes.

In selecting which of the duplicate responses would appear in this book, we settled on these two rules:

1. If the duplicates have identical wording, the earliest entry wins.
2. If not, the best wording wins.

Complaints about this system should be typed, single-spaced, printed on 8.5-by-11-inch paper, and then thrown away.

China Puts Lien on the State of California to Secure Their US Debt —@AskYourCoach

Martians Retreat after Discovering Earth's Natural Resources Are Depleted
—@rolfje

Auction Prices Paid for Paper Mache Art from the Last Century Hit a New High Due to Lack of Newspaper
—@garygoodenough

Gmail Out of Beta
—@girl_out_there

Fiat Chrysler Surpasses Tata GM in Global Auto Sales; Toyota Tesla Still #1
—@erictufts

First Resort Hotel Opens at South Pole—Year-Round Fun in the Sun! —@azcurt

Make up a likely headline from the year 2012.

Bill O'Reilly Calls for Boycott of World—Says Everyone a Pinhead —@rponto

World Does Not End; Mayans Bummed —@sketchler

Twitter Becomes Self-Aware, Destroys Self to Save Humanity —@ianeng

Obama Reelected in Reagan-style Landslide, 48–2 States Won —@SulaymanF

Twitter's Multibillion-Dollar Server Upgrade Complete; Messages Can Now Be 142 Characters —@shayman

Crackpot Theorists Proved Right: World Really Is En… —@fran6co

Microsoft's Latest 'iPhone Killer' Attains 8% Market Share —@bilmor

Twitter Buys Google, Divests Microsoft Division —@pmahoney87

Pogue Finally Exhausts 2009 Tweets, Releases "The World According to Twitter, Vol. 15." —@hriefs

Rupert Murdoch Admits Fox News Just an Out-Of-Control Bar Bet —@rponto

2012? No newspapers, no headlines. —@KirkAndrews

What's your million-dollar idea?

Biodegradable diapers with seeds sewn into them. Add the fertilizer, bury them in the ground. A few weeks later, flowers!
—@BruceTurkel

A service to tweet song requests to a coffee shop's music system. So I never hear Journey's "Faithfully" ever again.
—@rolando

Button on one sock, buttonhole on other. Use before washing them. Odd socks are history! —@vonschlapper

Combo business: Puppy day care & then rent the dogs out to guys who want to meet women.
—@tivoupgrade

Shoes that convert your kinetic energy into stored energy to charge your mobile device. —@mysterio_

140-calorie snacks to eat while Twittering. Call 'em Tasty Tweets. —@pumpkinshirt

An alcoholic beverage that's also a contraceptive. Called Plan Beer. Gonna be huge on campus. —@jesseburton

Software that you tell what to do with all your online accounts and info, once notified (by the county registrar?) that you are dead. —@Stefaniya

Oreo stuffing. Just by itself. In squeezable tubes. Like toothpaste. More sugar. Less mint. —@umapagan

Laptops with solar panels built in. —@tassoula

Open bakeries immediately adjacent to all 24-hour fitness locations, & vent the ovens into the HVAC system. —@BerryLowman

A dock for cellphone at main entrance that lets you answer calls, see SMS, etc. at extensions placed throughout a home—WANT! —@cpmomcat

Child-care and babysitting services and facilities at movie theatres. —@jessyz

A small device that makes your car emit the same space-age noises that George Jetson's did. Should be popular among hybrid owners. —@pumpkinshirt

An electromagnet that makes all the (magnetized) tennis balls on a court roll into the corner for easy pickup. —@SulaymanF

An email app with a 60-sec. delay on the Send key. Do nothing, it sends your outgoing email after 1 minute. You can abort during the minute. —@PeterWeisz

Jelly donuts...except instead of jelly, they're filled with soft-serve ice cream. It's a Sno-Nut. —@pumpkinshirt

A laser pruner: So you can fine-tune the view from any window. —@farinata

Universal sports helmet for kids—works for biking, skiing, hockey, baseball, etc.... —@reganref

TiVo + IMDB + face recognition so you can freeze-frame and find out where you know that guy playing the waiter from. —@pumpkinshirt

Mirrored contact lenses. —@automatt

Parents are run ragged by driving kids to/from school & activities. Run a livery service focused on this market. —@hriefs

Marriage meter: Premarital predictive compatibility device. —@DivorcingDaze

A toilet that analyzes your deposit & tells you what you need to be healthier (more fiber, drink more water, etc.). —@stains

Who's got the best boss-from-heaven story?

Boss gave me flowers + note when my work was rejected for publication. "People didn't like impressionism at first, either!" —@thinkc

I had a boss who'd hold team meetings at Hooters. All we ever did there was down beer and watch sports. —@snerra

Team had to travel to HQ over Valentine's Day. Boss sent every spouse a dozen roses in our name. —@tsmyther

"Tomorrow is another day. Let's pick this up then. Go enjoy the beautiful weather tonight." —@StefShoe

Trusted me to create, research and develop my job and title to help the company! Finally using my talents and degree :) —@SAngelSK

We listened to the "Weird Al" Yankovic song of the day at staff meetings. —@LifesizeLD

My boss pitched in to help cover my wedding costs. It was a sizable amount and totally unexpected! —@citygirlgvl

He let me leave in the middle of a very critical project to support a friend in trouble, and also advised me on handling the situation. —@aymroos

Great boss treated me to lunch at Keg steakhouse for performance evals. Had me order my first filet mignon. I was so poor. —@drewallstar

At the end of a rough week, a great boss told me, "Do something crazy this weekend, and bring me the receipt." —@Alli7on

We celebrate Beer O'Clock on Friday afternoons in the main conference room, per CEO's instructions. —@MelissaGaare

I was in the hospital recovering from an accident. Boss delivered donuts for nurses every day. Made me the nurses' favorite. —@denverandy

She gave up her raise because she felt that my raise was not enough, & there wasn't money in the budget to give me more. —@etreglia

My boss saw I was overworked (weeks of 18-hr days), & forced me to go home by unplugging my PC. Best sleep ever. —@jschechner

Me: Working and going to school, burning out, depressed. Boss: We'll give you six months' paid sabbatical so you can focus on school. —@bilmor

Mini fridge in office stocked with beer. And he shared. :) —@CafeChatNoir

Who's got the best boss-from-hell story?

Boss told others I took too many sick days. (I'd left work that day in an ambulance from a grand mal seizure!) —@Beppy

Boss used to tell us she was working from home—but had no Internet or even a computer in her apartment! —@janet1210

My boss fired me when I had to miss work one day because I felt sick from donating blood. —@i_Girl

My boss wouldn't give me the day off when my grandfather was having quad-bypass. Said: "What? Did he raise you or something?" —@BigDaddy978

A boss once demanded to know why I was 5 mins late at 9:05, even though I'd been called in at midnight (he called me in!). —@roybear

I had a boss who showed up late every Sat morning, still drunk, usually bloody from a fight... —@rstobie

My boss yelled for someone to come into his office. When she did, he said, at his PC: "How do I make the letters all slanty?" —@keithmarder

Got told that there was no work, as things were slow. Went in next week for final wage—there were 2 foreign nationals doing my old job. —@trekaddict

Directive from a boss: "No computers: Encourages bad habits"...It's a small business that still uses typewriters. —@drewbiondo

I had a boss who was born in a city called Hell, in Michigan. That was a true boss from Hell... —@CacauParazzoli

If I came late, my boss said: "Your hours are 9 to 5." If I left at 5, he'd say: "This is not a 9–5 job." —@victoryfarm

I was a politician's scheduler. He made me triple-book so he could pick the best gig—then I had to tell the rejects he wasn't coming. —@chaatiecakes

I had a boss who laid me off the day after my mom's funeral. I was out of the office, so had to have friends collect my things for me. —@tripodell

The boss' wife controlled payroll. When the boss didn't come home, sparks would fly, and paychecks would be held up to punish everyone. —@rogerhoward

Boss fired me in 1996 because I wouldn't come watch "Seinfeld" at his house with him and his wife. —@MPFOGARTY

Identify an irony of life.

Jalapeños go from green to red as they ripen, looking more hot, but actually becoming more sweet. —@_B_en

You shut down a Windows PC by clicking the Start button. —@Moonalice

As parents, we spend 18 years of love, time, and energy teaching our kids how to live successfully without us. —@cat8canary

The more rich and famous you are, the less you have to pay for things. —@kurometarikku

As a kid, you can't wait to be grown up so you can do anything you want. As an adult, you wish you were a kid for the same reason. —@sheppy

When you are young, you study in schools that you did not build: When you are old, you build schools in which you will never study. —@tsmyther

We spend the first 2 years of children's lives teaching them to walk & talk, and the next 16 telling them to sit down and shut up. —@nancilr

As a student, you have free time to travel but no money; after graduation, you have money but no time. —@jaimebatiz

The more someone likes/wants you, the faster you run for the hills! —@ascottfalk

Long-term customers (cable companies, cellphones) pay the highest prices. —@BetsC

Society's message to the young: Sex is a dirty, filthy, repulsive thing. Save it for the one you love. —@davidjan47

If something tastes really good, it's not good for you. —@satya893

Expending a lot of energy at the gym, even after having started out tired, makes you full of energy. —@wlerik

The less credit you need, the more offers you get. —@sfreedman

Sick people have to walk to the back of a drugstore to get medicine; healthy ones can buy cigarettes at the front counter. —@edamitz

Computers allow us to do things faster so we can spend more time sitting in front of computers wasting time. —@Sypher

What was your most memorable fortune cookie?

Definitely my oddest fortune ever (received shortly before I met my first husband): "Beware of strange men bearing used soda straws." —@pixelesh

"He who hesitates gets sideswiped." —@shyonelung

"He who takes last lettuce wrap will be least hungry." —@KeepAustinWired

One time, with my girlfriend, my fortune cookie said, "You will soon be pleasantly surprised." She opens hers—it says the same thing! —@trunks1022

Week before layoff: "You have an important new business development shaping up." On back: "Learn Chinese phrase: 'I need money.'" —@mudskip

"Send a message soon to the land of wind and ghosts." —@elpatro

I got this one in the '80s: "Do whatever you want today until the police catch up with you!" —@HeatherHAL

"Ignore previous cookie." —@stevehoff

"The next cookie fortune you get will be better than this one." (It wasn't.) —@NemaVeze

Pogue Sez

Usually, the process of asking nightly quesitons on Twitter went smoothly. Sometimes, though, my question wasn't clear, which confused the results.

This was one of them. What I actually tweeted was this: "TONIGHT'S BOOK QUESTION: Make up (or recall) a funny fortune-cookie fortune."

My followers responded with both real fortunes they'd encountered and fictional ones they thought would be funny.

In the end, the real fortunes were more entertaining, so, as you can see, I retroactively edited the question to refer to real fortunes. But a handful of the made-up fortunes were funny, too.

There were many submissions of the age-old, "Help! I'm being held prisoner in a fortune-cookie factory," of course. And there were many variations on, "That wasn't chicken."

But I also liked, "Nutrition Data. Total Fat 0g 0%, Cholesterol 0mg 0%, Sodium 22mg 1%, Total Carbohydrate 7g 2%, Fortune 200g 1%" (@iluvgs400). In a similar vein, there was, "Best used before 23 Apr 1985" (@rolfje).

Finally, I was reminded that this Twitter crowd is a computer-savvy one when @sheppy proposed this: "By opening this cookie, you accept the terms of the license agreement printed hereon..."

After I got engaged: "A nice cake awaits you."
—@hornsolo

"Soon. And in the presence of a young gentleman." (Yes, really.) —@Collegesandkids

"You will take a trip to the desert." (I haven't yet, and that was 5 years ago.)
—@saulyoung

Got one last weekend that said, "You are a happy man." (So far, it's 0 for 2.)
—@grimsb

My craziest fortune ever: "Here we go. 'Moo Shu Cereal' for breakfast with duck sauce."
—@mikemonello

"You look pretty."
—@bahamat

I got a fortune cookie once that read, "Strike while the iron is." Existential fortune!
—@pattyfab

I was 9 months pregnant, and my cookie said, "You'll meet a small stranger soon." For real! I still have it somewhere. —@Maggie_Dwyer

My favorite (real) existential fortune cookie: "Soon, and in good company." —@susanchamplin

My uncle once got, "You will get new clothes soon" as a cookie fortune. I was definitely jealous. —@annafran

No joke—a friend of mine got: "Try another fortune cookie." —@xtini

Real fortune: "Buy the red car." (Hubby was drooling over red 2010 Camaro for months...we ended up buying one!! They DO come true!) —@cstorms

"Reality is the only roadblock to hallucination." (Received that one last week!!) —@jamesus

From a Silicon Valley Chinese restaurant: "You will do well in the field of computer technology." (A safe fortune, but also funny!) —@donnaaparis

"The greatest danger may be your stupidity." (I got this fortune from a cookie during college. Still have it in my wallet.) —@drkent

"The person to your right will pay the check." (Father actually received...and was pleased.) —@hriefs

"There are always more fish in the sea; not as cute, nor as rich, but fish nevertheless." (Chinatown, Chicago 1986.)
—@jweinberger

"You are the crispy noodle in the vegetarian salad of life." (Absolutely real, tacked on the wall in front of me.)
—@andrewnathanson

"You will be hungry later. Order takeout now." —@brentfrederick

"You will be invited to a karaoke party." (I wasn't, and it's not much of a fortune!) —@aliceinthewater

"Be sincere, even when you don't mean it." —@elysea

"Winning numbers for Lottery 4, 24, 17, 16, 34 & 43, but you are a day late ordering dinner." —@BBWG

"It's a good thing that life is not as serious as it seems to the waiter." (Yes, we save our fortunes.) —@m0nkeyh0use

"Do not mistake temptation for opportunity." —@tsmyther

"Maybe the 'magic eight-ball' would suit you better."
—@AaronDWilliams

"You will find luck good." —@SusanEJacobsen

Make up the book title for a sure-fire bestseller.

The Big Book of Hundred Dollar Bills —@jschechner

Think and Grow Hair —@chaunceyc

How to Make Sense Out of the Books Meant to Help You Make Sense Out of Twitter —@mslavik

How to Retire on Little or No Money —@tammie423

Sleep Your Way Thin: Lose 7 lbs in 7 Hours —@vickyderd

Best Places to Live After Climate Change...And Their Schools —@matzzie

How to Save Your Retirement Fund Using Just Twitter —@andrewnathanson

Writing Books for Dummies for Dummies —@justcombs

Oprah Says to Buy this Book —@lizardrebel

So I Married a Ponzi Schemer —@giantsuberfan

How to Make Big Money Writing E-Books Telling People How to Make Big Money Writing E-Books —@BuckyKatt

YOU in a Threesome —@ikarl67

Boy, I Screwed Up! By G.W. Bush —@TomMendham

Where Obama Eats: An Annotated Guide to His Favorite Restaurants Across America —@wlerik

Define an ailment for the modern age.

Identical dissimilitude syndrome: A paralyzing fear in a clone that it is not like the other clones. —@AndyMaslin

Live-rticulitis: Affliction of sitting thru a TV program & commercials that haven't been TiVo'ed and trying to fast forward anyway. —@danielschwartz

Dropophobia: The fear of dropping the new $$$ gadget. Symptom: buying bulky cases that ruin the sleek design you loved and bought it for. —@nocturne1

Gizmodo syndrome: A disorder causing the patient to liveblog all major life announcements—weddings, christenings, funerals, etc. —@Notsewfast

TiVo jetlag: Inability to determine actual time or day of week, due to watching queued newscasts & programs out of sync. —@tasena

Cellular subsufficiency syndrome: Fear of being snubbed due to the cell phone you use being deemed pathetic. —@AndyMaslin

EMSADD: Excessive Media Stimulation Attention Deficit Dis... Oh! Cool video on YouTube! —@dhersam

Cleft politic: A deep divide between the right and left sides of society. —@ascottfalk

HALO3TOSIS: Shortness of breath caused by shouting obscenities into X-Box Live mouthpiece. —@UncJonny

Buy-polar disorder: An overwhelming urge to purchase beach-front property in the Canadian Arctic as a hedge against global warming. —@brianwolven

Reverse paranoia: You're afraid nobody's following you. —@kemulholland

Internetia significa: The condition of believing "because I can be Googled, I must matter." —@Brockeim

Bipolar idol syndrome: When you're torn between wishing you watched "American Idol" and having no desire to watch it at all. —@hornsolo

CWS (connectivity withdrawal syndrome): The datastream is interrupted and people become shaky, irritable, nauseous, then delusional. —@oreoteeth

PDADD: Attention-deficit disorder brought on by paying too much attention to handheld devices. —@CajunBrooke

Apatow syndrome: Mental disorder where one believes that despite ugliness, some wit + crazy adventure can win the girl. —@briguyd

Change 1 letter of a familiar title; explain.

There Will Be Brood: The Octomom story. —@Eshahan

Batch-22: Best homebrew ever. —@bilmor

A Beautiful Wind: Jennifer Connelly needs to get Russell Crowe some Gas-X. —@JimtheNinja

E.D.: The story of an aging alien with an embarrassing problem in the bedroom. —@reedkavner

Old Feller: Story about an endearing old man who eventually needs to be put down. —@dcwilhite

Angels & Lemons: Uncovering the secrets of the Desserted. —@BeelJDPhD

Dove in the Time of Cholera: And you thought avian flu was bad! —@claytonkroh

Par and Peace: The thickest Russian book ever on golf. —@lizardrebel

The Color of Monet: A has-been pool hustler leaves a bad impressionism. —@amy_harrison

Pilates of the Caribbean: Johnny Depp gives health class on tropical cruises. —@Bikkeltje

Bilk: Sean Penn plays a SF politician who pretends to be gay in order to defraud sympathetic voters to contribute to his campaign. —@jamesbritton

Allen: Sigourney Weaver meets a mildly interesting accountant from another planet. —@Notsewfast

Snapes on a Plane: Samuel L. Jackson can't stand the in-flight Harry Potter movie. —@dances_w_vowels

West Side Store: The gangs reconcile and realize the real money is in haute couture on 92nd and West End Ave. —@Sandydca

The B Team: A bunch of lazy old fat guys try their hardest to stop crime, but ultimately can't get a plan together. —@soneshk

Fetal Attraction: Doomed from the womb. —@mjantos

Independence Dad: Disgruntled father enlists alien allies in fight for male emancipation. —@4ntonChigurh

Iron Nan: Adventures of a cyborg granny. —@natesaint

Helpboy: Young demon joins the team at the IT help desk. —@brianwolven

Natal Attraction: Obsession and desire in the maternity ward. —@brianwolven

Scar Trek: Kirk recounts his space battle injuries for an adoring yeoman. —@brianwolven

Y-Men: Mutants who ask lots of questions. —@brianwolven

Bran Torino: Old man moves more than his car. —@garygoodenough

Ballets over Broadway: Woody Allen switches from mobsters to dancers. —@hughesviews

Binding Nemo: Clownfish signs exclusive multipicture deal. —@Greg_Bass

Lilo & Snitch: Enchanting story of a little girl and her blue tattletale friend. —@UberStated

Raging Gull: A beach bird gets really pissed off. —@smccormack

Schindler's Lisp: The story of a Nazi mocked for his speech defect. —@CRA1G

The Goofather: Thrilling account of the head of the Nickelodeon crime family. —@SKCOMEDY

Bleeper: Composed entirely of profane Woody Allen outtakes. —@eric_right_now

Apocalypse Wow: A journey into heart of darkness—with Riverdance! —@gametroll

Dickens' A Christmas Carob: Scrooge mends his ways after eating disappointing chocolate substitute. —@slightly99

The Whining: The story of a hotel where you don't have an axe to grind about bad room service. —@mrsseigleman

Moby-Sick: Ahab misses his big chance to catch the great white whale because he left his magnetic wristbands at home. —@mmarion

The Princess Bribe: Buttercup is offered millions to marry Humperdinck. —@grousehouse

Star Wart: Wherein we learn the real reason why Vader always wears that mask. —@brianwolven

What's the bright side of growing old?

Walker gives you a place to hang your Walkman.
—@jschechner

Always having the right-of-way while driving. Who dares cross your path? —@chelit

Four words: Get. Off. My. Lawn. —@pumpkinshirt

Hot nurse's aides. —@stickn317

People stop asking you to help them move.
—@oddthink

It's better than the alternative! —@ltb1014

You finally get to spend the fortune you socked away in your 401k... —@UncJonny

You stop sweating the small stuff. —@Aries419

You can use your cane like a light saber. —@chrisblake

Grandchildren. All the joy...none of the fuss. —@sdbeck

I no longer care if people think I'm odd. In fact, I prefer that they do. —@Dreadkiaili

You have finally figured out EXACTLY how you like your sandwiches. —@pumpkinshirt

Being able to say: "Wheeen I was youuur age..." —@marcelotrevino

You can watch the same movies a few months later, and enjoy them just as much! —@TiVoUpgrade

Caption this photo.

How long are these Doublemint commercial auditions gonna last? —@pumpkinshirt

With a quick flash on a somewhat meaningless day…3 children set off on an unknown journey. A journey into…The Twilight Zone. —@BoSSyBraT

Dolly the sheep seemed so much happier when we tested it on her. —@_B_en

You don't need Photoshop! Microsoft Paint can do it all. I'm 9 and I'm a PC. —@justcombs

Aw, Dad, do we HAVE to do the von Trapp thing every time company is over? —@pumpkinshirt

Captain Kirk, sir, I think we're having a wee bit o' trouble with the transporter. Again. —@joe221

Tonight on "60 Minutes": Imaginary friends. The truth revealed. —@interweave

Wish the four girls from "The Shining" would come over to play already. —@robtruman

Take your time with the lineup, ma'am. They can't see you. Which one is the perpetrator? —@Hollenbach

That's it? That's all LSD does? Stupid hippies! —@ElePhatt

For the fourth time, honey, I don't think we need the fertility drugs. —@mldrabenstott

Little did the kids know, their past selves would haunt them for the rest of their lives. —@pathumx

Who let the clones out? —@CharPrincessa

As you get closer to the black hole, the gravitational pull increases. Further down, it takes two hands just to hold up your head. —@wgseligman

Our future selves and present selves can never meet, or the space-time continuum will collapse...Oh, damn. —@JC_zoracel

Alliances were formed early on in Survivor: Daycare. —@alechosterman

Scientists, Parents, Children Discover Cloning Not Nearly As Exciting As Predicted. —@cat8canary

Stupid shutter lag! —@madscienti11

What's the most you ever spent for a meal?

A $600 sushi dinner for two in Tokyo was fantastic until I got the bill. It was in yen and I figured it out back at the hotel. Ouch! —@davidjan47

A $900 dinner for two at Everest in Chicago to ring in the year 2000. I'll scale back when the next millennium rolls around. —@hriefs

A little north of $1,000 at The French Laundry. Just because. Oh, and $2 for a pickle at JFK Terminal 6. —@myny28

Around $3,000 for my son's college grad dinner: 4 adults, 1 child. I bribed him—we'd take him to Tru if he graduated in 4 yrs. A bargain! —@ChicagoDiane

$1,500 for five people at Guy Savoy. Cool experience but $1,500 is more than I pay for rent. Felt a bit guilty. —@ritam

$50. Because I've never had enough money to spend more. —@tjbliss

I once paid $120 for dinner for two after promising the waitress a 100% tip if she got a complicated order just right. She did. —@sheppy

$600 for two, omakase at The Hump. Memorable! (But the fois gras sushi with gold flakes was a bit much.) —@krisztinaholly

About $2,200 for 4. Sashimi & Sake to celebrate Super-Computer acceptance at TIT in Tokyo, Japan. —@egreshko

Once got a $250 bonus that had to be spent at a restaurant. Wife and I went to a fancy place and lived it up for just over $250. —@yacitus

$27 for a hot dog from a Vatican vendor, 'cause I didn't know the difference between 1,000,000 lire & 100. —@kimanastasia

10th anniversary, 7-course chef's menu with wine at El Bizcocho: $400 for 2. More performance than meal, worth every cent. —@Stefaniya

Split a $203 meal for two because it was my friend's b'day and her hubby had just left her—and I couldn't afford even my half. —@ascottfalk

$400 Aureole, Vegas 1998. Getting married the next day. Wife won $500 on a slot and still had enough to get her hair done. —@finnadat

Pogue Sez

Here's another one of those questions that left the wrong impression on Twitter newbies.

As I retweeted some of the best responses, I used the standard syntax: "RT @barney" (meaning, "I'm re-broadcasting Barney's response").

But beginners who didn't know what RT meant thought that these were all *my* expensive-meal stories.

"Wow, dude, you must be a trillionaire," wrote one.

Well, not really. But thanks for the thought.

$262 for fixed-price Xmas dinner for two. We are both vegetarian, so they removed the meat. That's big $ for sauteéd veggies! —@DrewJazz

I spent $350 buying steak dinners for my in-laws...after I won on "Jeopardy!" —@lavasusan

We spent $303.33 for dinner on our 1st anniversary. Seven years later, we had triplets. Glad the bill wasn't $404.44! —@tjredbird

Took a cab from Newark to John's Pizza (NYC) and back during a layover. Pizza: $19. Taxi fares: $110. Total craving: $129. —@DrewJazz

$600 to take a Tibetan lama's widow & her assistant to one of the best SF restaurants, in gratitude for a school her husband founded. —@stevesilberman

$785. My brother & I had invited family & mom's friends to a favorite restaurant in lieu of a memorial service—but many, many more came. —@lsmith1964

$220, to impress my girlfriend. Sadly, she broke up with me the day after. —@imamathwiz

Several hundred $-lunch at Le Jules Verne in the Eiffel Tower. Hubby promised it if I survived kidney cancer surgery. —@pamheld

$214 about 9 years ago. She said yes. —@johnmarkharris

What should your vanity license plate be?

Bright red Yaris: HABANERO —@jnaz

On a Camaro: I8 A 4RE —@Vanish

NE1410S? —@PoopTheWorld

VNTY PLT —@benjamin_gray

STOPTXTING —@Vanish

IXLR8 —@ksalemi

SHOPAHOLIC —@Starry9281

BACHROX —@rponto

N8 WZDM —@pumpkinshirt

BACK OFF —@mikemaughmer

1337 R1D3 —@Game_Kid

S5280FEET —@CMich

DRAWKCAB —@pumpkinshirt

DRIVER 404 —@hriefs

URNAMEHERE —@msewall

HANGUPNDRIV —@SALYGO

YBNRML —@lakelady

ENGINERD (and proud to admit it) —@taminas

NEWJOBPLS —@dawns8

HONKAGAIN —@holle8e3

TWEET ME —@cmumathwhiz

I82MUCH —@pumpkinshirt

1LSTR8R —@mordy

PAYS2ADV —@GBLPR

HMSTR PWRD —@DrewJazz

T1 3VOM (reads "move it" in rearview mirror) —@davegiroux

ST0L3N —@XmasRights

3MORPYMNTS —@tomtwine

IO-2WRKIGO —@tsmyther

ILUVCOPS —@smccormack

READL8R —@pumpkinshirt

MMMBACON —@sarahbeery

2SXY4MYCAR —@yoshionthego

BADSPLR —@pumpkinshirt

NCC-1701-F —@MisterGlass

You can send one tweet back in time. You can give advice but not reveal events to come. What and who?

To: Me. Date: 2005. Subject: Buy Apple and Google stocks. —@janzimmermann

Donner party: Don't forget the trail mix. —@rponto

To JFK: Mr. President, I think you should consider not using a convertible today. It's sunny, but you never know, it might rain... —@gustaborja

To Steven Spielberg three years ago: Don't listen to George Lucas. Indiana Jones does not need a new entry into the franchise. —@mehughes124

Message to myself (ca. 2002): Do not, I repeat do not, accept a job offer while running a 103°F fever. —@mccoma

1907 to Vienna Academy of Fine Arts: You really should reconsider rejecting A. Hitler's application. He has great artistic potential! —@rachelij

Gore, 2001: Someday you're gonna look back and laugh...really hard. —@cpearsall

Dear newspaper editors, circa 1995: Post your news online. Charge for it. Don't give away the ads, either. —@meranduh

To Cinderella: Don't forget to tell the prince your name. —@meranduh

To: Senior class of 1988: Lose mullet before class photos. —@pumpkinshirt

ca. 1912: Honestly, Captain Smith, I think guests on the maiden voyage would appreciate a slightly more southerly route. —@Curious_Objects

November 2000, Florida: Let's use the electronic voting machines. —@hriefs

To: Young George Washington: Brush and floss regularly. —@pumpkinshirt

Mr. President, let's cross-check that WMD intel... —@VenetianBlond

To Romeo: Don't drink the poison right away. Savor the moment. —@arielleeve

Neo: Take the blue pill. —@mikefeigin

ca. 1950: Mr. Madoff, you're very good at gaining people's trust... Have you considered becoming a rabbi? —@Notsewfast

Janet, why don't you wear an extra bra underneath that bustier... It's the Super Bowl and you can't be too careful! —@Notsewfast

TO: Myself, 10 minutes ago: Stop after 2 donuts. TO: Myself, 5 minutes ago: Stop after 3 donuts. —@pumpkinshirt

You can send one tweet back in time. You can give advice but not reveal events to come. What and who?

To my boyfriend, September 1983: Don't forget the condoms! —@smccormack

Mr. Lincoln, the play's terrible! Take the missus out to a nice dinner instead. —@Gene_in_FL

To Adam: When Eve offers you the apple, say NO! —@ravelez

Mr. Bush, you should increase your ownership in the TX Rangers. Really, baseball is where you belong. —@TheBubbaTex

You've lived your life this far. What have you learned?

You are your own worst enemy. Fortunately, you can be defeated. —@jlawrencem

If you can see the top of the fridge, you have to clean the top of the fridge. (With great power comes great responsibility.) —@UberStated

My 50% Rule: Spend half your energy expressing yourself; spend the other half being quiet, open to everything around you. —@rponto

Allowing someone to fail can be the best way to teach them to succeed. —@AndyMaslin

Dance more. —@johnmsilva

It's ALL dubious. —@Jim_OConnell

You can fix anything but a broken heart. —@JC_zoracel

Older men and younger technology are the way to go. —@tassoula

Life's like a roll of toilet paper. Plenty of time when you start, but goes SO fast as you get nearer the end. —@tanabutler

Even when you're sure it's about you, it's not about you. —@erinvang

Never judge by the smile. Always look at the eyes.
—@Tymethief

Don't sweat the petty things, and don't pet the sweaty things. (George Carlin) —@theshorneagle

If you just say things exactly as they are, people think you're funny. —@LifesizeLD

It is either too late or too early to worry.
—@fieldtripearth

Life is not a zero-sum game. But you shouldn't assume the other guy knows that, too. —@ToddABalsley

"Beloved Father & Husband" is a better epitaph than "Beloved Employee." —@vonschlapper

You can't make anyone love you. —@LibraryBarbara

Life does not have an upward trajectory, but is rather a series of ups and down, often toward a beneficial end. —@EShahan

Ideas are cheap. Execution is all that matters. —@techsutra

Just be nice and respectful to others. Squeaky wheels may get grease, but little else. —@jasonact

Take aim, THEN fire. —@hriefs

There's *always* something new to learn. It's the journey, not the destination. —@trishm

Never break up with someone during the movie's opening credits. —@pumpkinshirt

Don't wait for company to use the good china. —@kstew

New Yorkers aren't rude. You're just walking too damn slow. —@myhelfy

When your children become adults, your love for them is no less intense than when they were babies. —@SharonZardetto

Hang onto the ones you love, fiercely. They're going to decide what home you go into. —@rpnrch

There are 13 symmetrically arranged holes in a saltine cracker. Even an odd number can be beautifully balanced. —@hughesviews

I've learned to ask. Surprising what is given to those who ask. —@kellycroy

Pick the one person or thing in life most important to you, and schedule your day around it. —@cswriter

Figure out what you like to do—then find someone to pay you to do it. —@sressler

The more we learn, the less we know. —@bilmor

If it is not important in 30 days, it's not important. —@Jimg

Never play cards or pool with anyone who has a colorful nickname. —@pumpkinshirt

Family comes first. —@grousehouse

The average American knows the price of everything and the value of nothing. Don't be average. —@DrKoob

You only get one chance to see your kids grow up. —@mdsharp

Steer clear of pessimists. They'll only drag you down and they can't be changed no matter how hard you try. —@MacRtst

Don't date your ex. —@gametroll

Never watch a new TV show. If it's good enough to last, catch it on DVD. —@Wildecat101

Righty tighty, lefty loosey. —@medicwhite

Everyone eventually turns into their parents, though for some reason are still surprised when it finally happens to them. —@David_Minton

Don't regret failing. Because 9 times out of 10, you end up learning more from it than succeeding in the first place. —@meancode

The older I get, the smarter my father appears to be. —@markinark

You can't change the direction of the wind, but you can adjust your sails. —@rodkoerner

It doesn't matter how fabulous or important I might think I am; at the end of the day, I've still got to clean up after the cats. —@hannnahkl

Things change. Always. —@TheOnlineMom

Every bad day contains the seeds of a better one. —@ajbezark

Practice doesn't make perfect—it makes me late for the other things I need to do. —@SusanEJacobsen

When the system works against you, make a new system. —@theatermonkey

Very few things are as important as we see them in the moment. —@batogato

Wasabi is NOT the same as guacamole, no matter how similar they look. —@dndgirl

No matter how many new clothes you buy, you still end up wearing the same 8 outfits. —@AnnieLaG

You CAN have it all, just not necessarily all at once. —@BeNewmann

The smell of bacon cheers up everyone in the house. —@grabbingtoast

I've learned enough to fill 140 bookshelves, not 140 characters. —@bpdobson

You know you're a Twitter addict when...

...u knw hw 2 abbrv evry wrd n nglsh. —@_wendy_r_

...you've planned your last words to be 140 letters or less. —@danblondell

...you're happy to hit a red light so you can safely check your phone while driving. —@Stefaniya

...you're able to play Scrabble even though most of the vowels are missing.

—@pumpkinshirt

...ur grammr resembles that of teen8ger. —@kstew

...you create & constantly update a Twitter account for your dog—and the dog has more followers than you do! —@lvdjgarcia

...guests in your living room catch you tweeting when you escape to your bathroom. You know who you are... —@carpathia16

...you're in rock bands named The Fail Whalers, Twick or Tweet, Corned Beef Hashtags, or One Hundred Forty and the Characters. —@pumpkinshirt

...you have to be put into a medically induced coma every time Twitter has a planned outage for maintenance. —@pumpkinshirt

...you begin speaking words using Tw- at the start. E.g., I like to twavel by twain. —@FastFoodMom

...your 11-year-old son joins Twitter so that he can talk to you. —@CharJTF

...you run into a chatty neighbor & ask, "Could I get the 140-character version?" —@RosanneLambert#oops

...you assume no one around you but the person you're talking to can hear you if you start your conversations with "D." —@Vanish

...your wife says she can't sleep because the birds are tweeting outside. You say "unfollow them" and go back to sleep. —@pumpkinshirt

...you worry that Guy Kawasaki isn't posting often enough. —@ajbezark

...you are momentarily confused by a URL that doesn't point to bit.ly. —@cakatz

...you started on the easy stuff like MySpace, graduated to snorting Facebook, and now are lying in an alley, mainlining tweets. —@SKCOMEDY

...your friends try to check you into the Maureen Dowd Twitterholic Recovery Center. —@phofland

...you lose your job at Twitter for excessive at-work tweeting. —@pumpkinshirt

...you're reading and tweeting via iPhone while your wife is having your baby at 2 a.m. (like I did Wed night). —@dszp

...you click the Block button to silence your children—but for some reason they can still speak to you. —@christopherbmac

...you ask your lawyer to draw up a new will dividing your estate evenly among your 2,419 followers. —@pumpkinshirt

...you use the word "tweeple" without irony. —@kawika

...you take available Twitter usernames into account when naming your child. —@ImageDistillery

....your children have to call you "@mom" when they speak to you. —@christopherbmac

...instead of air quotes, you find yourself making "air hash-marks" in conversation. —@pumpkinshirt

... you say "at" before your friends' actual names. —@awwsboss

...you ask to be buried with your smartphone...just in case. —@pumpkinshirt

...you create another account just to follow your primary account. —@ehphotography

...your family says so. —@bmark42

Index of Contributors